T0213603

The aim of this Series is to highlight the latest international findings and advances in plant pathology and plant disease management, and plant pathology topic specialist, Congress and Workshop organisers, coordinators of broad International projects are invited to consult with the Series Editor regarding their topic's potential inclusion in the series.

Under an agreement with Springer, a Book Series based on the invited lectures at the 9th International Congress of Plant Pathology ICPP2008, *Plant Pathology in the 21 st Century* was initiated and four books covering key themes were published. Following on the procedure for the 2008 Congress, three additional volumes in the series were published on themes which were key topics at ICPP2013, held in Beijing, China, while two new books are almost ready from ICPP 2018 held in Boston (SA). Moreover, more books were published in between Congresses, covering up-to-date topics in plant pathology. In light of the initial seven volumes' success, the ISPP has now concluded an agreement with Springer to broaden the scope of the Series and publish additional volumes.

The goal of the International Society for Plant Pathology (ISPP; www.isppweb. org) is to promote the global advancement of plant pathology and the dissemination of essential information on plant diseases and plant health management. This book Series looks of particular interest due to the upcoming International Year of Plant Health (2020).

Ana Vučurović • Nataša Mehle
Géraldine Anthoine • Tanja Dreo • Maja Ravnikar
Editors

Critical Points for the Organisation of Test Performance Studies in Microbiology

Plant Pathogens as a Case Study

Editors
Ana Vučurović
Department of Biotechnology and Systems
Biology
National Institute of Biology
Ljubljana, Slovenia

Géraldine Anthoine
French Agency for Food, Environmental
and Occupational Health and Safety
(ANSES), Plant Health Laboratory
Angers, France

Maja Ravnikar
Department of Biotechnology and Systems
Biology
National Institute of Biology
Ljubljana, Slovenia

Nataša Mehle
Department of Biotechnology and Systems
Biology
National Institute of Biology
Ljubljana, Slovenia

University of Nova Gorica, School for
Viticulture and Enology
Vipava, Slovenia

Tanja Dreo
Department of Biotechnology and Systems
Biology
National Institute of Biology
Ljubljana, Slovenia

European Union's Horizon 2020 research and innovation programme
Funded in the frame of Valitest project (grant agreement n°773139)

(cc) (i) BY

ISSN 2512-160X   ISSN 2512-1626   (electronic)
Plant Pathology in the 21st Century
ISBN 978-3-030-99813-4   ISBN 978-3-030-99811-0   (eBook)
https://doi.org/10.1007/978-3-030-99811-0

© The Editor(s) (if applicable) and The Author(s) 2022. This book is an open access publication.
Open Access This book is licensed under the terms of the Creative Commons Attribution 4.0 International
License (http://creativecommons.org/licenses/by/4.0/), which permits use, sharing, adaptation,
distribution and reproduction in any medium or format, as long as you give appropriate credit to the
original author(s) and the source, provide a link to the Creative Commons license and indicate if changes
were made.
The images or other third party material in this book are included in the book's Creative Commons license,
unless indicated otherwise in a credit line to the material. If material is not included in the book's Creative
Commons license and your intended use is not permitted by statutory regulation or exceeds the permitted
use, you will need to obtain permission directly from the copyright holder.
The use of general descriptive names, registered names, trademarks, service marks, etc. in this publication
does not imply, even in the absence of a specific statement, that such names are exempt from the relevant
protective laws and regulations and therefore free for general use.
The publisher, the authors and the editors are safe to assume that the advice and information in this book
are believed to be true and accurate at the date of publication. Neither the publisher nor the authors or the
editors give a warranty, expressed or implied, with respect to the material contained herein or for any
errors or omissions that may have been made. The publisher remains neutral with regard to jurisdictional
claims in published maps and institutional affiliations.

Cover illustration: Virus-like disease symptoms on tomato fruit infected with tomato spotted wilt
orthotospovirus, cucumber mosaic virus and southern tomato virus. Photograph by Nejc Jakoš
(National Institute of Biology, Ljubljana, Slovenia).

This Springer imprint is published by the registered company Springer Nature Switzerland AG
The registered company address is: Gewerbestrasse 11, 6330 Cham, Switzerland

Foreword

The recent COVID-19 pandemic permitted to better understand, if still necessary, the importance of rapid, precise and reliable detection of pathogens. Such a need is common in all cases: plants, animals and humans. Probably, pathogen detection is one of the major fundaments of circular health, a concept becoming more and more popular after the COVID-19 pandemic. Under this perspective, the book "Critical points for the organisation of test performance studies in microbiology: plant pathogens as a case study" is touching a very much up-to-date topic.

Regulated or non-regulated plant pests (bacteria, virus, fungi, nematodes, arthropods or weeds) are responsible of the major crop losses. Their accurate and reliable detection and identification are essential to avoid or reduce the economic costs and trade disruption, and to support surveillance activities.

This book is based on the results achieved within the Horizon 2020 VALITEST Project, aiming at producing validation data for plant pathogens detection, also working with numerous tests based on new technologies developed to meet the different needs. The book describes well throughout its chapters the process of harmonising practices, with two rounds of test performance studies. The first round includes combinations of pest/test/matrix, prioritised based on the expertise of the Project consortium. The second round includes other combinations based on the needs expressed by various stakeholders. Appropriate statistical approaches and the use of technologies, such as next-generation sequencing, permitted to improve the harmonised procedures in plant health. Liaison with regional and international standardisation bodies has facilitated large dissemination of validation data obtained, especially, by their inclusion in harmonised diagnostic protocols. The approaches adopted to maximise the impact of the Project are described, including the validation programme suppliers outside the consortium, and allow the participation to the test performance studies of voluntary proficient laboratories.

This book will hopefully allow, by reaching a larger public, to further strengthen the interactions between stakeholders in plant health for better diagnostics, thus providing the agricultural industry one of the most important components of sustainable plant protection.

Agroinnova, University of Torino,   Maria Lodovica Gullino
Grugliasco, Italy

Preface

The ongoing coronavirus (COVID-19) pandemic has demonstrated the critical importance of rapid and reliable pathogen detection. Similarly, plant production is frequently endangered by outbreaks of different pests. The recent emergence of several pests, such as *Xylella fastidiosa* and tomato brown rugose fruit virus, has required rapid responses from all stakeholders in the field of plant health, including diagnostic laboratories, to allow national plant protection organisations to take appropriate measures based on a reliable diagnosis. The increased international trade in plants and climate change are frequently resulting in the expansion of pest distributions, which as a consequence is putting enormous pressure on agro-ecosystems and the stakeholders involved in food production to provide enough food for the growing human population. Therefore, the use of timely and coordinated measures is essential to avoid losses and to achieve sustainable plant-based food production as well as prevent and limit damage caused by plant pests. Successful pest control will become even more important in the future in the context of further reductions in the use of chemical pesticides.

Guidelines for plant pest diagnosis are provided in the International Standard for Phytosanitary Measures 27 'diagnostic protocols for regulated pests', developed by the International Plant Protection Convention (IPPC). In addition, guidelines on diagnostics are also developed by Regional Plant Protection Organizations (RPPOs). The European and Mediterranean Plant Protection Organization (EPPO) is the RPPO for the Euro-Mediterranean region, and it has had an established work programme on diagnostics since the 1990s and has developed a number of horizontal and pest-specific diagnostic standards (series PM 7 of EPPO Standards).

A variety of tests that have been developed by commercial companies or research institutions are used to diagnose pathogens. However, the reliability of diagnostic tests depends on the intended use of the tests, their performance characteristics, and the associated uncertainty obtained from validation studies and the experience of the laboratories. Diagnostic laboratories ensure that a test is fit for purpose through the validation process, which is based on the evaluation of its performance characteristics, such as analytical sensitivity, analytical specificity (inclusivity and exclusivity), selectivity, repeatability and reproducibility. The performance characteristics of a test can be determined within one laboratory (i.e., 'in-house') during a validation study (referred to as intralaboratory studies) or by several laboratories in interlaboratory comparisons. Guidance on the organisation of interlaboratory comparisons is provided in the EPPO Standard PM 7/122(1). There are different types of interlaboratory comparisons, but this book focuses on the organisation of test performance studies (TPS) that are aimed at the evaluation of the performance of a test, or tests, by two or more laboratories using defined samples. The need for reliable diagnosis in plant health has been recognised by the European Commission and was supported by the funding of the VALITEST Project (www.valitest.eu, 2018–2021; grant agreement N° 773139) through the EU Horizon 2020 research and innovation programme.

The purpose of this book is to provide practical and technical guidance for the organisation of TPS for plant pests based on the experience gained by the TPS organisers through the organisation of 12 TPS in the framework of VALITEST. The main aspects and challenges of preparing, organising and reporting TPS are highlighted and designed to be of help to the organisers of future TPS not only in the field of plant pest detection but also in other areas of microbiology. The possibility to use the knowledge gathered in this book beyond the field of plant health will enable the creation of new network connections and exchanges, and as an outcome, this will continuously improve the concept and organisation of TPS. We hope that this book will help to increase the number of laboratories that are willing to organise TPS for the global benefit of the plant health diagnostic community.

We would like to express our gratitude and appreciation to all our colleagues who have accepted to be co-authors of the chapters in this book. As this book is part of the Springer book series Plant Pathology in the Twenty-First Century, we are thankful to Prof Dr Maria Lodovica Gullino (series Editor), and Zuzana Bernhart and her group at Springer for their kind support and help during the preparation of this book.

Ljubljana, Slovenia

Angers, France

Ana Vučurović
Nataša Mehle
Géraldine Anthoine
Tanja Dreo
Maja Ravnikar

Acknowledgements

The authors would like to thank Jenny Tomlinson and Lynn Laurenson from Fera Science Ltd. (FERA) for their help during the preparation of this book.

We are thankful to all of the colleagues who participated in the VALITEST Project, to all of our colleagues who provided their isolates for the TPS organisers, and to all of the laboratories that took part in the 12 TPS organised in the framework of the VALITEST Project.

We would like to specifically thank Sébastien Massart and Bénédicte Lebas (University of Liège, ULG) for useful discussions on the composition of the sample panel with regard to statistical analysis of the data, and René van der Vlugt (Wageningen University & Research) for preparation of the guidelines on the reference materials.

We also thank our colleague Dr Maja Zagorščak from NIB for her help with R codes for the Heatmap.

Contents

1 General Background 1
Françoise Petter, Charlotte Trontin, Géraldine Anthoine, Nataša Mehle,
Maja Ravnikar, Tanja Dreo, Tadeja Lukežič, and Ana Vučurović

2 Introduction to Interlaboratory Comparisons 7
Françoise Petter, Charlotte Trontin, Géraldine Anthoine,
Maja Ravnikar, Tanja Dreo, Tadeja Lukežič, Ana Vučurović,
and Nataša Mehle

3 Description of the Process of TPS Organisation 15
Géraldine Anthoine, Ian Brittain, Anne-Marie Chappé,
Aude Chabirand, Tanja Dreo, Francesco Faggioli, Catherine Harrison,
Nataša Mehle, Monica Mezzalama, Hanna Mouaziz,
Tom M. Raaymakers, Jean-Philippe Renvoisé, Marcel Westenberg,
Françoise Petter, Charlotte Trontin, Tadeja Lukežič, Ana Vučurović,
and Maja Ravnikar

4 Conclusions 61
Ana Vučurović, Géraldine Anthoine, Charlotte Trontin, Tanja Dreo,
Tadeja Lukežič, Françoise Petter, Maja Ravnikar, and Nataša Mehle

Appendices 65

References 95

Contributors

Géraldine Anthoine French Agency for Food, Environmental and Occupational Health and Safety (ANSES), Plant Health Laboratory, Angers, France

Ian Brittain Animal and Plant Health Agency (APHA), Sand Hutton, York, UK
Detection and Surveillance Technologies, Fera Science Ltd., Sand Hutton, York, UK

Aude Chabirand French Agency for Food, Environmental and Occupational Health and Safety (ANSES), Unit for Tropical Pests and Diseases, Saint Pierre, Reunion Island, France

Anne-Marie Chappé French Agency for Food, Environmental and Occupational Health and Safety (ANSES), Nematology Unit, Le Rheu, France

Tanja Dreo Department of Biotechnology and Systems Biology, National Institute of Biology, Ljubljana, Slovenia

Francesco Faggioli Research Centre for Plant Protection and Certification, Council for Agricultural Research and Economics, Rome, Italy

Catherine Harrison Detection and Surveillance Technologies, Fera Science Ltd., Sand Hutton, York, UK

Tadeja Lukežič Department of Biotechnology and Systems Biology, National Institute of Biology, Ljubljana, Slovenia

Nataša Mehle Department of Biotechnology and Systems Biology, National Institute of Biology, Ljubljana, Slovenia
University of Nova Gorica, School for Viticulture and Enology, Vipava, Slovenia

Monica Mezzalama Centre of Competence for the Innovation in the Agro-environmental Field (AGROINNOVA), University of Torino, Grugliasco, Torino, Italy

Hanna Mouaziz French Agency for Food, Environmental and Occupational Health and Safety (ANSES), Angers, France

Françoise Petter European and Mediterranean Plant Protection Organization, Paris, France

Tom M. Raaymakers National Reference Centre for Plant Health, Dutch National Plant Protection Organization, Netherlands Food and Consumer Product Safety Authority (NVWA), Wageningen, The Netherlands

Maja Ravnikar Department of Biotechnology and Systems Biology, National Institute of Biology, Ljubljana, Slovenia

Jean-Philippe Renvoisé French Agency for Food, Environmental and Occupational Health and Safety (ANSES), Quarantine Unit, Lempdes, France

Charlotte Trontin European and Mediterranean Plant Protection Organization, Paris, France

Ana Vučurović Department of Biotechnology and Systems Biology, National Institute of Biology, Ljubljana, Slovenia

Marcel Westenberg National Reference Centre for Plant Health, Dutch National Plant Protection Organization, Netherlands Food and Consumer Product Safety Authority (NVWA), Wageningen, The Netherlands

Chapter 1
General Background

Françoise Petter, Charlotte Trontin, Géraldine Anthoine, Nataša Mehle ⓘ**, Maja Ravnikar** ⓘ**, Tanja Dreo, Tadeja Lukežič, and Ana Vučurović** ⓘ

1.1 Introduction

The ongoing coronavirus (COVID-19) pandemic has caused a global health crisis, which has demonstrated the critical importance of rapid and reliable pathogen detection. Similarly, in the field of plant health, the recent emergence of several pests in the EPPO region, such as *Xylella fastidiosa* and tomato brown rugose fruit virus, required rapid responses from diagnostic laboratories and other stakeholders to allow national plant protection organisations to take appropriate measures based on a reliable diagnosis.

The movement of pests has considerably increased in the past century because of increased and diversified international trade. Climate change also has an impact on plant health, for examplethe expansion of pest distribution (IPPC 2021). In addition,

F. Petter (✉) · C. Trontin
European and Mediterranean Plant Protection Organization, Paris, France
e-mail: petter@eppo.int; trontin@eppo.int

G. Anthoine
French Agency for Food, Environmental and Occupational Health and Safety (ANSES), Plant Health Laboratory, Angers, France
e-mail: geraldine.anthoine@anses.fr

N. Mehle
Department of Biotechnology and Systems Biology, National Institute of Biology, Ljubljana, Slovenia

University of Nova Gorica, School for Viticulture and Enology, Vipava, Slovenia
e-mail: natasa.mehle@nib.si

M. Ravnikar · T. Dreo · T. Lukežič · A. Vučurović
Department of Biotechnology and Systems Biology, National Institute of Biology, Ljubljana, Slovenia
e-mail: maja.ravnikar@nib.si; tanja.dreo@nib.si; tadeja.lukezic@nib.si; ana.vucurovic@nib.si

© The Author(s) 2022
A. Vučurović et al. (eds.), *Critical Points for the Organisation of Test Performance Studies in Microbiology*, Plant Pathology in the 21st Century 12, https://doi.org/10.1007/978-3-030-99811-0_1

the rapid growth of the human population and the resulting increased demand for food are putting enormous pressure on agro-ecosystems and the stakeholders involved in food production (Rodrigues et al. 2017). To avoid losses and achieve sustainable plant-based food production, prevention or limitation of damage caused by plant pests through the use of timely and coordinated measures is essential. The rapid and accurate detection of plant pest is a foundation of successful pest control and will become more important in the future in the context of further reductions in the use of chemical pesticides.

Guidelines for plant pest diagnosis are provided in the International Standard for Phytosanitary Measures 27 'diagnostic protocols for regulated pests' developed by the International Plant Protection Convention (IPPC). IPPC Standards are recognised by the World Trade organization as reference Standards for international trade. In addition, guidelines on diagnostic are also developed by Regional Plant Protection Organisations (RPPOs). The European and Mediterranean Plant Protection Organization (EPPO) is the RPPO for Europe,[1] and has established a work programme on diagnostics since the 1990s, and developed a number of horizontal and pest specific diagnostic Standards (series PM 7 of EPPO Standards). Additional requirements may be specified by other national bodies involved in plant health.

A variety of tests are used to diagnose pathogens. These tests may be developed by commercial companies or research institutions. The reliability of diagnostic tests depends on the intended use of the tests, their performance characteristics and associated uncertainty obtained from validation studies and the experience of the laboratories (EPPO PM 7/76 2018a). Validation, i.e., the process by which laboratories ensure that a test is fit for purpose based on the evaluation of its performance characteristics, consists of several steps described in Fig. 1.1, each of which needs to be carefully planned and executed. The most important part of the validation process is the determination of the performance characteristics of a test. Performance characteristics that are frequently used to characterise tests include: analytical sensitivity, analytical specificity (inclusivity and exclusivity), selectivity, repeatability and reproducibility (EPPO PM 7/98(5) 2021a). Guidance for the evaluation of these performance characteristics has been developed by EPPO and is included in the Standard PM 7/98(5) (2021a). The performance characteristics of a test can be determined within one laboratory (i.e., 'in-house') during a validation study (referred to as intralaboratory studies) or by several laboratories in interlaboratory comparisons. Guidance on the organisation of interlaboratory comparisons is provided in the EPPO Standard PM 7/122(1) (2014). There are different types of interlaboratory comparisons, but this book focuses on the organisation of test performance studies which aim at evaluating the performance of (a) test(s) by two or more laboratories using defined samples. Overall, validation is a demanding process in terms of time and resources (expertise, money). In plant health, given the large number of pests, matrices and methods and the combinations of these, data on the performance of

---

[1] Beyond Europe members of the Organization are from the Mediterranean Area and also Central Asia.

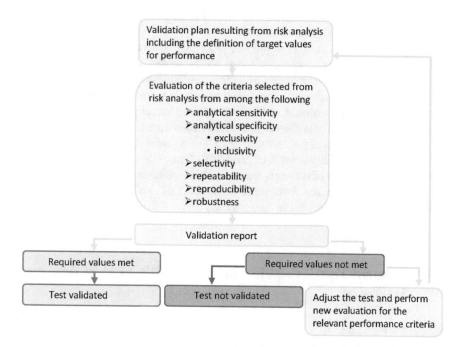

Fig. 1.1 The validation process in Plant Health. (Adapted from EPPO PM 7/98 (4) (2019))

diagnostic tests is not always available and validation of tests is mainly performed by laboratories based on their need to be accredited ISO/IEC 17025 (2005) to perform their activities.

The need for a reliable diagnosis in plant health has been recognised by the European Commission and was supported by the funding of the VALITEST Project (www.valitest.eu, 2018–2021; grant agreement N° 773,139) through the EU Horizon 2020 research and innovation programme. The most important aims of the VALITEST Project were to: (1) provide more complete and precise descriptions of the performances of diagnostic tests; (2) stimulate, optimise and strengthen the interactions between stakeholders in plant health to promote better diagnostics; and (3) lay the foundations for structuring the quality and commercial offers of plant health diagnostics tools. Fulfilling these goals was ensured by the creation of a multidisciplinary consortium that brought together leading EU public, private, academic and industrial organisations and other stakeholders from the plant health and diagnostics sectors.

One of the core activities of the Project was the production of validation data for existing tests through the organisation of 12 TPS. In total the performance of 83 tests covering 11 pests recognised as a priority was evaluated during two rounds of TPS (Trontin et al. 2021). As the provision of validation data for selected tests is time consuming and requires a significant investment of human and financial resources, one of the outcomes of VALITEST was also the development of a guidance documents to assist in the organisation of TPS, and to improve the diagnostic procedures and the validation framework.

The purpose of this book is to provide practical and technical guidance for the organisation of TPS for plant pests based on the experience gained by the TPS organisers through the organisation of TPS in the framework of VALITEST. The major aspects and challenges faced during the preparation, organisation and reporting of TPS are identified and can be used by organisers of future TPS not only in the field of plant pest detection, but also in other areas of microbiology. They are mainly illustrated using the case study of a TPS organised on tomato spotted wilt orthotospovirus in the framework of VALITEST but this example can be easily adapted to the specifics of different tests for which validation is required. Practical templates developed in the framework of VALITEST for specific steps of TPS organisation are also given in the Appendices.

1.2   Common Terms Used in This Book (EPPO PM 7/76 (5) (2018a))

Accuracy - the ability of a test to detect true positives and true negatives, as (true positives + true negatives)/total population of negatives.

Analytical sensitivity - the smallest amount of the target that can be detected reliably, and sometimes referred to as the 'limit of detection'. Further details on the procedures to determine analytical sensitivity are given in EPPO PM 7/98 "Specific requirements for laboratories preparing accreditation for a plant pest diagnostic activity".

Analytical specificity (including inclusivity and exclusivity; see below) - further details on the procedures to determine analytical specificity (inclusivity, exclusivity) are given in PM 7/98 "Specific requirements for laboratories preparing accreditation for a plant pest diagnostic activity".

Diagnostic sensitivity - the proportion of infected or infested samples that test positive, as compared with results from alternative tests (or a combination of tests).

Diagnostic specificity - the proportion of uninfected or uninfested samples that test negative (i.e., true negatives) compared with results from alternative tests (or a combination of tests).

Exclusivity - performance of a test with regards to cross-reactions with a range of non-targets (e.g., closely related organisms, contaminants).

Inclusivity - the performance of a test with a range of target organisms that covers genetic diversity, different geographic origins or hosts.

Interlaboratory comparison - organisation, performance and evaluation of measurements or tests on the same or similar items by two or more laboratories in accordance with predetermined conditions (i.e., proficiency testing, test performance studies).

Matrix - type of material (e.g., leaves of tomato, pepper seeds ...).

Methods - methods include bioassay methods, biochemical methods, fingerprint methods, isolation/extraction methods, molecular methods, morphological and morphometric methods, pathogenicity assessments and serological methods.

Repeatability - the level of agreement between replicates of a sample tested under the same conditions.

Reproducibility - the ability of a test to provide consistent results when applied to aliquots of the same sample tested under different conditions (e.g., different times, operators, equipment, locations).

Robustness of a test - the extent to which altered test conditions (e.g., temperature, volume, reagents) affect the established test performance values (e.g., analytical sensitivity, analytical specificity).

Selectivity - the extent to which variations in the matrix affect the test performance (i.e., matrix effects).

Test - the application of a method to a specific pest and a specific matrix.

Test performance study (ring tests, collaborative trials) - evaluation of the performance of one or more tests by two or more laboratories using defined samples (evaluation of a test).

Validation - process carried out to provide objective evidence that a test is suitable for the circumstances of its use (ISO/IEC 17025 2005).

Open Access This chapter is licensed under the terms of the Creative Commons Attribution 4.0 International License (http://creativecommons.org/licenses/by/4.0/), which permits use, sharing, adaptation, distribution and reproduction in any medium or format, as long as you give appropriate credit to the original author(s) and the source, provide a link to the Creative Commons license and indicate if changes were made.

The images or other third party material in this chapter are included in the chapter's Creative Commons license, unless indicated otherwise in a credit line to the material. If material is not included in the chapter's Creative Commons license and your intended use is not permitted by statutory regulation or exceeds the permitted use, you will need to obtain permission directly from the copyright holder.

Chapter 2
Introduction to Interlaboratory Comparisons

Françoise Petter, Charlotte Trontin, Géraldine Anthoine, Maja Ravnikar (ID), Tanja Dreo, Tadeja Lukežič, Ana Vučurović (ID), and Nataša Mehle (ID)

2.1 Different Types of Interlaboratory Comparisons

There are two types of interlaboratory comparison studies: proficiency tests, which aim at monitoring the proficiency of laboratories and test performance studies (TPS; or collaborative method validation studies, collaborative trials, ring tests), which aim at evaluating the performance of (a) specific test(s) and whether it (they) is (are) fit for purpose (ISO 17025 2005, EPPO 2014 PM 7/122(1)). The main differences between TPS and proficiency tests are shown in Fig. 2.1. In this booklet only TPS organisation is covered.

A TPS is usually organised to monitor the performance of a newly developed test to detect and/or identify 'emerging' pests or strains of known pests or to compare the performance of different tests. The results of TPS also provide information on how

F. Petter (✉) · C. Trontin
European and Mediterranean Plant Protection Organization, Paris, France
e-mail: petter@eppo.int; trontin@eppo.int

G. Anthoine
French Agency for Food, Environmental and Occupational Health and Safety (ANSES), Plant Health Laboratory, Angers, France
e-mail: geraldine.anthoine@anses.fr

M. Ravnikar · T. Dreo · T. Lukežič · A. Vučurović
Department of Biotechnology and Systems Biology, National Institute of Biology, Ljubljana, Slovenia
e-mail: maja.ravnikar@nib.si; tanja.dreo@nib.si; tadeja.lukezic@nib.si; ana.vucurovic@nib.si

N. Mehle
Department of Biotechnology and Systems Biology, National Institute of Biology, Ljubljana, Slovenia

University of Nova Gorica, School for Viticulture and Enology, Vipava, Slovenia
e-mail: natasa.mehle@nib.si

© The Author(s) 2022
A. Vučurović et al. (eds.), *Critical Points for the Organisation of Test Performance Studies in Microbiology*, Plant Pathology in the 21st Century 12, https://doi.org/10.1007/978-3-030-99811-0_2

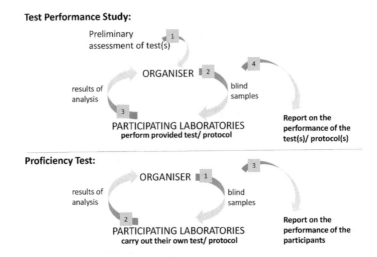

Fig. 2.1 Main differences between test performance studies and proficiency tests in plant health

(a) test(s) perform(s) in different laboratories; i.e., on different equipment, with different reagents, different personnel. This allows a better estimation of the accuracy and reproducibility of tests. The organisation of a TPS is a complex process that requires significant efforts from the organisers in terms of time, expertise and finances. A TPS may be organised following a request from different stakeholders involved in plant health, such as National Plant Protection Organisations (NPPOs), companies producing commercial tests, and diagnostic laboratories, as part of their activities.

TPS organisation includes different steps that are connected and are mutually dependent (Fig. 2.2). The required steps in TPS organisation are sometimes sequential, while others can occur simultaneously. The first two steps are presented in Sects. 2.2 and 2.3 of this chapter whereas the other steps are presented in Chap. 3.

2.2 Selection of Pests

For the validation of test(s) for a specific pest a choice should be made by the laboratory to organise a TPS or to perform an intralaboratory validation. The selection of the pests for which a TPS should be organised is important considering resources needed for validation processes (e.g., the personnel, financial requirements, biological material, etc.). The aim of selection is to direct resources to critical points and/or where improvement of tests has the highest beneficial impact. Ideally, test selection should involve different stakeholders and reflect current and potential future needs in plant health. Pests for which there is a need for an appropriate test are selected taking into account their current importance or their potential to have a

Selection of pests

Selection of the TPS organizers

Selection of the tests for TPS

- Definition of the TPS scope and selection of diagnostic methods
- Definition of weighted criteria and targeted value for each criterion
- Collection of available validation data
- Definition of the list of tests subjected to preliminary studies
- Assessment of the results of preliminary studies against weighted criteria

Selection of the TPS participants

- Identification of potential participants
- Definition of weighted criteria for the selection
- Invitation of potential participants
- Analyses of eligibility of potential participants

Contracts and technical information for TPS participants

Preparation and dispatch of samples and reagents

- Definition of the test panel composition (reference material/ samples included in TPS)
- Stability and homogeneity testing (samples and reagents)
- Dispatching

Collecting TPS results and analyzing data

Reporting of TPS results

Fig. 2.2 Steps involved in test performance study organisation

significant impact in the near future. The expression of a need to conduct a TPS may come from different stakeholders. The VALITEST Project focused on pathogens and partners selected 11 pests (including viruses, bacteria, nematodes and fungi) that were of interest to stakeholders in the region. The first six target pests for the first round of TPS were selected during the preparatory phase of the Project, by matching a list of candidate pests with the experience of the VALITEST partner laboratories. This ensured that laboratories with expertise in specific pests organised the relevant TPS. The pests that were selected for this first round are listed in Table 2.1.

For the second round of the VALITEST TPS, the targets were selected based on stakeholder and market needs and identified through two online surveys by two online surveys launched by EPPO at the beginning of the Project (Agstner and Jones 2020). For the second round of the VALITEST TPS, the targets were selected based on stakeholder and market needs (Trontin et al. 2021). At the beginning of the Project, two online surveys were launched for the list of pests categorised by the EU, one for laboratories and one for NPPOs. Respondents to both surveys were asked to

Table 2.1 Pests and associated test performance study organisers selected for the first round of Test performance studies in the framework of the VALITEST Project

Pest	Pest group	EU pest status (Commission Implementing Regulation (EU) 2019/2072)	Test performance study organiser
Erwinia amylovora	Bacteria	Protected zone quarantine pest (annex III), regulated non-quarantine pest (annex IV)	NIB
Pantoea stewartii subsp. *stewartii*	Bacteria	A1 quarantine pest (annex II A)	NIB
Citrus tristeza virus	Virus	A1 quarantine pest (annex II A), protected zone quarantine pest (annex III), regulated non-quarantine pest (annex IV)	ANSES[a]
Bursaphelenchus xylophilus	Nematode	A2 quarantine pest (annex II B)	ANSES[b]
Plum pox virus	Virus	Regulated non-quarantine pest (annex IV)	NVWA
Fusarium circinatum	Fungus	A2 quarantine pest (annex II B)	FERA

NIB: National Institute of Biology, Ljubljana, Slovenia; [a]ANSES: Plant Health Laboratory, French Agency for Food, Environmental and Occupational Health and Safety, Unit for Tropical Pests and Diseases, Saint Pierre, Reunion island, France; [b]ANSES: Plant Health Laboratory, French Agency for Food, Environmental and Occupational Health and Safety, Nematology Unit, Le Rheu, France; NVWA: Dutch National Plant Protection Organisation, National Reference Centre, Wageningen, The Netherlands; FERA: Fera Science Ltd., York Biotech Campus, Sand Hutton, York, United Kingdom

indicate their current testing priorities. The laboratories were instructed to rank the pests according to the number of tests performed. NPPOs were instructed to rank the pests based on current plant health priorities. In the second survey representatives of NPPOs were instructed to rank their top 10 pests based on current plant health priorities. The surveys were distributed through the EPPO networks, and the results were combined for the final selection of the pests. The six pests selected for the second round of TPS are listed in Table 2.2.

2.3 TPS Organiser

The capacity (in terms of both staff expertise and availability) of the TPS organiser is essential to ensure the reliability of the results of such study. The requirements for the organisation of interlaboratory comparisons are described in EPPO PM 7/122 (1) 2014 and ISO 17043 (2010). TPS organisers should have appropriate technical and scientific competence in relation to the organism and the test(s) to be performed, they should remain impartial, and should maintain confidentiality throughout the process. In practice this means that an external reader of a TPS report should not be able to identify which results were obtained by which participant (unless the participant waives confidentiality). When a laboratory commits to organise a TPS, this should be carefully planned and communicated with the interested stakeholders.

Table 2.2 Pests and associated test performance study organisers selected for the second round of test performance studies in the framework of the VALITEST Project

Pest	Pest group	EU pest status (Commission Implementing Regulation (EU) 2019/2072)	Comments for prioritisation for validation	Tests commercially available	Test performance study organiser
Tomato brown rugose fruit virus	Virus	Emergency measures	Major concerns for growers of tomato and pepper	Yes	CREA
Tomato spotted wilt orthotospovirus	Virus	Regulated non-quarantine pest (annex IV)	Harmful for ornamental plants, vegetables and industrial crops	Yes	NIB
Plum pox virus (on site testing)	Virus	Regulated non-quarantine pest (annex IV)	Important for fruit tree certification; new strains emerging.	Yes	ANSES[a]
Xanthomonas citri pv. *Citri*	Bacteria	A1 quarantine pest (annex II A)	Major concerns for citrus leaves, stems and fruit; not present in the EU	Yes	ANSES[b]
Xylophilus ampelinus	Bacteria	Regulated non-quarantine pest (annex IV)	Major pest on grapevine; impact on trade and EU exports	Yes	FERA
Cryphonectria parasitica	Fungi	Protected zone quarantine pest (annex III), regulated non-quarantine pest (annex IV)	Major concerns for chestnut and other susceptible tree species	Yes	UNITO

CREA: Consiglio per la Ricerca in Agricoltura e l'Analisi dell'Economia Agraria (CREA), Centro di Ricerca Difesa e Certificazione, Rome, Italy; NIB: National Institute of Biology, Ljubljana, Slovenia; [a]ANSES: Plant Health Laboratory, French Agency for Food, Environmental and Occupational Health and Safety, Quarantine Unit, Lempdes, France; [b]ANSES: Plant Health Laboratory, French Agency for Food, Environmental and Occupational Health and Safety, Unit for Tropical Pests and Diseases, Saint Pierre, Reunion island, France; FERA: Fera Science Ltd., York Biotech Campus, Sand Hutton, York, United Kingdom; UNITO: AGROINNOVA - Centre of Competence University of Torino, Grugliasco, Italy

Laboratories organising TPS should carefully prepare the plans and determine a feasible time frame, they need to take into account unforeseen problems. Some empirical examples of how certain unforeseen events can affect the time frame are given in Chap. 3. A laboratory organising a TPS should define a person responsible for the organisation of this TPS, and should ensure the availability of personnel, equipment and facilities needed for the organisation. The organiser should have a quality assurance system in place (ISO 17025 (2005) or equivalent). A TPS organiser needs to keep records of all the activities and to provide traceability of the measurement results.

2.3.1 Selection of TPS Organisers

Test performance studies can be organised individually or, in some cases, in batches (i.e., more than one TPS at the same time, organised by the same or different organisers or consortium) when necessary. When several laboratories are part of a network a choice may need to be made of which one should organise the TPS. For the first round of TPS organised in the framework of VALITEST, TPS organisers were selected based on laboratories' expertise with a selected list of pests (Table 2.1). For the second round of TPS (Table 2.2), TPS organisers were selected based on the results of an online poll, where each potential organiser expressed their interest to organise one or more specific TPS on a preselected list of prioritised pests. Organisers judged their expertise regarding the list of pests. When more than one laboratory expressed an interest in organising a TPS for one pest, the TPS organiser was selected based on predefined criteria that were transparent and approved by all the parties involved. These criteria are described below.

2.3.2 Minimum Requirements for TPS Organisers

A list of minimum criteria that should be met by any TPS organiser was defined and is detailed below. A TPS organiser must also fulfil all requirements that are prescribed for TPS participants (see Sect. 3.3.2 "Definition of weighted criteria for the selection of TPS participants"). In addition, it is an advantage if a TPS organiser has experience in organising similar interlaboratory comparisons.

2.3.2.1 Involvement in Diagnostic Activities with Different Methods

A TPS organiser needs to have significant experience in the performance of diagnostic activities (e.g., performing routine analyses) using a range of different methods and matrices (e.g., type of plant material: seed, leaves, etc.) for the pest in question or in the case of an emerging pest for a pest of the same genus or family.

2.3.2.2   Experience with the Development and Validation of Diagnostic Tests

The personnel in the laboratory organising a TPS needs to demonstrate continued experience in the validation of diagnostic tests in their field, and to have adequate knowledge, competence and experience. Experience in the development of diagnostic tests can be considered as an additional advantage. Records of staff qualifications and training should be available, for example through a quality management system.

2.3.2.3   Experience with the Shipment of Material

Materials for TPS often require special shipping conditions. These can include special requirements for shipping on dry ice or for shipping certain quarantine pests, for which participants might need to provide specific documentation, such as a Letter of Authority. Consequently, it is important that a TPS organiser has experience in shipping similar materials abroad. Completing the official paperwork for shipping specimens according to Plant Health Regulations can take time, which should be considered by the TPS organiser when planning the TPS.

2.3.2.4   Collaboration with National and International Bodies in the Field of Plant Pest Diagnostics

It is preferable that TPS organisers are part of diagnostic networks at national, regional (e.g., EPPO) or international (e.g., IPPC) level or participate in Euphresco. Participation in such networks facilitates collaborations within the field, ensures an efficient flow of information, and provides an opportunity to communicate and disseminate TPS results to the target audience, thereby increasing the impact of the Project.

2.3.2.5   Technical Requirements

A TPS organiser should use agreed terminology and processes, have an established quality assurance system, be able to prepare reference material, and ensure that the validation processes are properly documented. In the EPPO region, terminology is defined in PM 7/76 (EPPO 2018a) and processes for validation defined in PM 7/98 (EPPO 2021a). The use of harmonised terminology can prevent misunderstandings. It is also an advantage if a TPS organiser has already participated in at least one interlaboratory comparison study, preferably in a TPS.

Open Access This chapter is licensed under the terms of the Creative Commons Attribution 4.0 International License (http://creativecommons.org/licenses/by/4.0/), which permits use, sharing, adaptation, distribution and reproduction in any medium or format, as long as you give appropriate credit to the original author(s) and the source, provide a link to the Creative Commons license and indicate if changes were made.

The images or other third party material in this chapter are included in the chapter's Creative Commons license, unless indicated otherwise in a credit line to the material. If material is not included in the chapter's Creative Commons license and your intended use is not permitted by statutory regulation or exceeds the permitted use, you will need to obtain permission directly from the copyright holder.

Chapter 3
Description of the Process of TPS Organisation

Géraldine Anthoine, Ian Brittain, Anne-Marie Chappé, Aude Chabirand,
Tanja Dreo, Francesco Faggioli, Catherine Harrison, Nataša Mehle ⓘ,
Monica Mezzalama, Hanna Mouaziz, Tom M. Raaymakers,
Jean-Philippe Renvoisé, Marcel Westenberg, Françoise Petter,
Charlotte Trontin, Tadeja Lukežič, Ana Vučurović ⓘ,
and Maja Ravnikar ⓘ

Supplementary Information The online version contains supplementary material available at
[https://doi.org/10.1007/978-3-030-99811-0_3].

G. Anthoine (✉)
French Agency for Food, Environmental and Occupational Health and Safety (ANSES), Plant
Health Laboratory, Angers, France
e-mail: geraldine.anthoine@anses.fr

I. Brittain
Animal and Plant Health Agency (APHA), Sand Hutton, York, UK

Detection and Surveillance Technologies, Fera Science Ltd., Sand Hutton, York, UK
e-mail: ian.brittain@apha.gov.uk

A.-M. Chappé
French Agency for Food, Environmental and Occupational Health and Safety (ANSES),
Nematology Unit, Le Rheu, France
e-mail: anne-marie.chappe@anses.fr

A. Chabirand
French Agency for Food, Environmental and Occupational Health and Safety (ANSES),
Unit for Tropical Pests and Diseases, Saint Pierre, Reunion Island, France
e-mail: aude.chabirand@anses.fr

T. Dreo · T. Lukežič · A. Vučurović · M. Ravnikar
Department of Biotechnology and Systems Biology, National Institute of Biology, Ljubljana,
Slovenia
e-mail: tanja.dreo@nib.si; tadeja.lukezic@nib.si; ana.vucurovic@nib.si; maja.ravnikar@nib.si

F. Faggioli
Research Centre for Plant Protection and Certification, Council for Agricultural Research and
Economics, Rome, Italy
e-mail: francesco.faggioli@crea.gov.it

© The Author(s) 2022
A. Vučurović et al. (eds.), *Critical Points for the Organisation of Test Performance
Studies in Microbiology*, Plant Pathology in the 21st Century 12,
https://doi.org/10.1007/978-3-030-99811-0_3

3.1 The TPS Organisation Process

Organising a TPS involves different steps that are inter-connected. The steps are mostly sequential, but some may be conducted simultaneously, such as *Selection of the tests for TPS* (parts dedicated to preliminary studies) and *Selection of the TPS participants* (see Fig. 3.1). This chapter details the following: the steps regarding the selection of the tests to be validated (Sect. 3.2); the selection of the laboratories to participate in the TPS (Sect. 3.3); the preparation of the materials and the dispatch of the samples (Sect. 3.5); and the completion of the TPS (including the collection and analysis of the TPS results) (Sect. 3.6). There is the need to plan the appropriate number of samples (including replicates) and of laboratories that should be included in the TPS to ensure an appropriate statistical analysis. Based on the experience from the two rounds of TPS in the VALITEST Project, the expected time for the completion of a TPS is approximately 1 year. An example of a Gantt chart (designed for a TPS organised in the framework of the VALITEST Project) is given in Fig. 3.1. It is worth noting that the TPS presented in the Gantt chart was organised within strict time frames, and certain steps could not be prolonged due to time limitations

C. Harrison
Detection and Surveillance Technologies, Fera Science Ltd., Sand Hutton, York, UK
e-mail: catherine.harrison@fera.co.uk

N. Mehle
Department of Biotechnology and Systems Biology, National Institute of Biology, Ljubljana, Slovenia

University of Nova Gorica, School for Viticulture and Enology, Vipava, Slovenia
e-mail: natasa.mehle@nib.si

M. Mezzalama
Centre of Competence for the Innovation in the Agro-environmental Field (AGROINNOVA), University of Torino, Grugliasco, Torino, Italy
e-mail: monica.mezzalama@unito.it

H. Mouaziz
French Agency for Food, Environmental and Occupational Health and Safety (ANSES), Angers, France
e-mail: hanna.mouaziz@anses.fr

T. M. Raaymakers · M. Westenberg
National Reference Centre for Plant Health, Dutch National Plant Protection Organization, Netherlands Food and Consumer Product Safety Authority (NVWA), Wageningen, The Netherlands
e-mail: t.m.raaymakers@nvwa.nl; m.westenberg@nvwa.nl

J.-P. Renvoisé
French Agency for Food, Environmental and Occupational Health and Safety (ANSES), Quarantine Unit, Lempdes, France
e-mail: jean-philippe.renvoise@anses.fr

F. Petter · C. Trontin
European and Mediterranean Plant Protection Organization, Paris, France
e-mail: petter@eppo.int; trontin@eppo.int

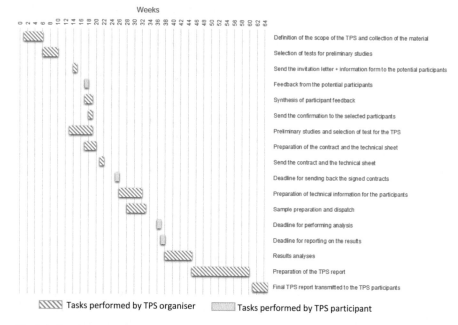

Fig. 3.1 Example of a Gantt chart for the organisation of a test performance study in the framework of the VALITEST Project (the duration of the steps can be extended in other contexts)

connected to the duration of the Project. When organising a TPS outside of a specific project, it is worth considering a longer time frame for the organisation. Steps that could require more time than in the Gantt chart example are: selection of test for preliminary studies (in particular when many tests are available for the pest of interest, as was the case for TSWV), and preliminary studies and result analysis.

3.2 Selection of the Tests

The first step of the TPS consists of the selection of the tests to be evaluated in the TPS. The importance of this step depends on the available pest tests and the amount of available validation data for each test, plus expert knowledge on the particular pest. It is often not possible to validate all tests for a selected pest due to limited resources, are therefore criteria need to be defined (depending on the needs at the time), and tests need to be selected based on the most important ones. Unbiased analysis of the available data on the performance characteristics of the tests will enable informed decisions to be made on the selection of the most relevant tests for the TPS. During the process, the scope of the TPS needs to be kept in mind, while also considering the resources available, including the budget, staff, equipment and materials.

The first step of the process is the definition of the purpose and scope of the TPS. The second step is the definition of the weighted criteria and associated target values, to facilitate the selection of the tests to be included in the TPS. The use of weighted criteria allows an impartial evaluation of the validation data available. Once the criteria have been defined, it is possible to proceed with a comprehensive analysis of the validation data available. When validation data are lacking, tests may be selected for preliminary validation studies, which are conducted in the laboratory of the TPS organiser by authorised personnel supervised by the person responsible for the TPS organisation. After completion of the preliminary studies, the organiser analyses the results and selects the tests to be included in the TPS based on the previously defined criteria.

The result of the test selection process is a list of tests selected to be included in the TPS. These tests have extensive intralaboratory validation data, although an evaluation involving several laboratories has never been performed. Throughout the TPS, all of the procedures are documented, and records are maintained in accordance with the quality assurance system of the laboratory organising the TPS.

3.2.1 Definition of the TPS Scope

To be able to set the strategy and to define the priorities for the test selection, it is necessary to precisely define the scope of the TPS based on the aim of the study. The definition of the scope includes the selection of the methods that will be used, and for each method, the identification of: sample type (e.g., DNA, sample spiked with pest), matrix (e.g., seeds, leaves), purpose (e.g., detection, identification), controls, number of samples and maximum number of participants. The selection of the methods depends on the diagnostic needs for the pest(s). In some cases, tests are needed for fast on-site detection, while in other cases, detection of the pest at low levels is more important. Methods differ in terms of their reliable detection of pests in symptomatic or asymptomatic materials. In the framework of the definition of the TPS scope, the selection of the methods might also depend on the plant material available and the expertise of the TPS organiser.

Defining the scope of the TPS is an important step in TPS organisation, as it will impact on all the other steps. The scope of the TPS should be clearly defined regarding the aim of the test(s), and its (their) feasibility, as the majority of TPS have limited resources and tight time schedules. Table 3.1 provides an example from VALITEST to illustrate how the scope of the TPS was defined for the detection and identification of tomato spotted wilt orthotospovirus (TSWV) in symptomatic leaves of tomato (*Solanum lycopersicum* L.) (Table 3.1), using both serological (ELISA) and molecular (RT-PCR, real-time RT-PCR) methods. In addition, some tests were evaluated for their applicability for on-site use. The other scopes of the TPS organised in the framework of VALITEST were detailed in the deliverable reports of the Project (for details see Alič et al. 2020; Anthoine et al. 2020).

Due to time constraints, it was not possible to use infected plant material from tomato in the VALITEST TSWV TPS. As a result, the starting material included extracts of healthy tomato leaves spiked with different virus isolates at various concentrations. TSWV isolates are very diverse in terms of the host plant in which they are detected, geographic origin, and molecular and serological properties. Consequently, determination of the analytical specificity of the tests was considered essential. Different TSWV isolates (i.e., geographic, biological) were used to ensure that the tests selected covered the majority of the known TSWV isolates (to evaluate

Table 3.1 Scope definitions for a test performance study for tomato spotted wilt orthotospovirus

Scope	Methods			
	ELISA[a]	Methods applicable for on-site	RT-PCR	Real-time RT-PCR
Sample type (DNA, plant material with deactivated pests, etc.)	*Infected/ non-infected plant material*	*Infected/ non-infected plant material*	*Infected/ non-infected plant material*	*Infected/ non-infected plant material*
Matrix (type of plant material: Seed, leaves, etc.)	*Leaves of tomato*	*Leaves of tomato*	*Leaves of tomato*	*Leaves of tomato*
Suitable for: Symptomatic / asymptomatic sample	*Symptomatic*	*Symptomatic*	*Symptomatic*	*Symptomatic*
Purpose: Detection / identification	*Detection and identification*	*Detection and identification*	*Detection and identification*	*Detection and identification*
Type of controls needed (NIC, NAC, PAC, PIC, IC, etc.)	*PC, NC*	*PC, NC*	*PAC, PIC, NAC, NIC*	*PAC, PIC, NAC, NIC, IC*
Number of samples	*22*	*22*	*22*	*22*
Maximum number of participants	*20*	*20*	*20*	*20*

PC, positive control; NC, negative control; NIC, negative isolation control; PIC, positive isolation control; PAC, positive amplification control; NAC, negative amplification control; IC, internal control

[a]*italics* text, needs to be determined by the TPS organiser

the analytical specificity [inclusivity]). The analytical specificity (exclusivity) of the selected tests was determined in the preliminary studies by including different orthotospovirus species similar to TSWV based on their serological and molecular properties. Usually, the main constraints in the organisation of the TPS are the availability of biological material, the availability of personnel and funding resources, and the limited time frame to conduct the TPS. Taking into account those constraints, the TPS organiser considered it feasible to include 22 samples and the relevant controls, following the guidelines described in Massart et al. (2022), to be analysed by approximately 20 participating laboratories. It is of utmost importance that all methods or tests that are planned to be included in a TPS are well established in the laboratory of the TPS organiser, and that appropriately trained or experienced personnel are available to perform and supervise the process.

3.2.2 Definition of Weighted Criteria and Targeted Values for the Selection of Tests

Usually, the number of tests available to detect a specific pest is significantly higher than the number of tests that can be included in a TPS, except for emerging pest for which the number of tests can be low. Not all available tests for a given pest are suitable for a TPS, and selection of the suitable tests should be impartial and transparent, and organised to achieve the best possible results with the resources available. Therefore, it is best to first define the criteria that have to be used for the selection of the tests. The list of criteria that can be used for the selection of tests may include the performance characteristics of the tests that are important for a particular intended use, the experience of the organiser, and the applicability of the test (see applicability, chemistry, instrument . . .). In addition, there is the need to define the type of value a criterion can take (e.g., quantitative, qualitative), called the criteria descriptors below, the targets to be reached by the test, and the relative weight of each criterion. The weights applied can be different for different uses of a test (e.g., laboratory, on-site). The weighted criteria need to be set to objectively select tests from a list of tests for a specific pest, each of which will have advantages and disadvantages depending on the scope of the TPS.

Criteria descriptors can be quantitative; e.g., the concentration of a pest that needs to be detected (not all pests can be counted easily; e.g., viruses). Descriptors can also be simple Yes/No answers, or relative levels. As already emphasised above, the criteria can be weighted differently to allow the selection of the appropriate tests for the defined scope of the TPS. The most important criteria are considered first (i.e., those with high weight); if some of the tests show similar values and performances, then less important criteria can be used as well (i.e., for those with medium or low weight). The important performance criteria for tests for diagnostic purposes are related to the following performance criteria: analytical sensitivity, analytical specificity (i.e., exclusivity, inclusivity), selectivity, repeatability and reproducibility. Other criteria can help to evaluate other properties of specific tests, such as applicability, reagents and equipment. Sometimes, when selecting between test with similar characteristics, a test with a higher sample throughput and easier test procedures can have an advantage. However, it is important to also evaluate how accessible or stable the required reagents are, and the equipment that is needed to perform a specific test for a specific scope. If a criterion is not relevant for a specific method or pest combination, it can be ignored (i.e., not given any weighting).

The values of the weights that are assigned to the criteria can differ depending on the scope of the TPS. For example, for emerging pests, the aim of the TPS might be to select tests to detect the pest at low concentrations in asymptomatic plant material (e.g., in seeds). Conversely, for testing of symptomatic material, detection at low concentrations is not critical.

The list of criteria defined here is not specific to TPS nor to plant health, and can be used for intralaboratory studies as well as for TPS in other fields. These criteria can thus be considered as the common rules for the selection of tests included in such studies, whereby they can be adapted to specific purposes when needed. Table 3.2 lists these common rules for the selection of tests for validation, which need to be defined and described to ensure the transparency of the selection process (for details see Alič et al. 2020; Anthoine et al. 2020). The TPS organiser should also define targeted values to be reached by a test for each criterion.

Table 3.2 is an example of how the targeted values and weights were defined for each criterion for a TPS organised for the detection and identification of TSWV. Here, it was very important that the test chosen detects all of the correct targets (i.e.,

Table 3.2 Criteria for the selection of tests for the test performance study for tomato spotted wilt orthotospovirus

Criterion	Descriptor	Target	Relative weight	
			Laboratory	On-site
Validation data (prior preliminary studies)				
Is the target (gene/protein) appropriately selected?	Yes/no	Yes[a]	High	High
Available validation data	Yes/no	Yes	Medium	Medium
Validation data available for selected matrix	Yes/no	Yes	Medium	Medium
Analytical sensitivity (LOD) (pure culture or DNA diluted in water)	Concentration (absolute, if possible; relative; low/medium/high)	Low RNA dilution	Medium	Low
Analytical sensitivity in plant material (selected matrix)	Concentration (absolute, if possible; relative; low/medium/high)	Low dilution of plant material	Medium	Low
Diagnostic sensitivity (comparison of different tests)	%	Within the same method: More sensitive test	Medium	Low
Analytical specificity	Level	Specificity: TSWV only	High	High
(a) Exclusivity (non-target organism): False positives	Level	0%	High	High
(b) Inclusivity (target organisms): False negatives	Level	0%	High	High
Selectivity	Presence of cross-reactions with matrix	No	High	High
Repeatability (near LOD)	Level	100% at LOD	Medium	Medium
Reproducibility/ robustness	%	100% at LOD	Medium	Medium
Results of interlaboratory comparisons available	Yes/no	Yes	Low	Low
Additional information (not a criterion!)				
Type of matrix				
Extraction method				
Use on symptomatic/ asymptomatic material				
Other Used successfully in different laboratories (according to literature) Part of the target genome				

(continued)

Table 3.2 (continued)

Criterion	Descriptor	Target	Relative weight	
			Laboratory	On-site
Validation data (after preliminary studies)				
Analytical sensitivity (LOD) (pure culture or DNA diluted in water)	Concentration (absolute, if possible; relative; low/medium/ high)	*Low dilution of RNA*	Medium	Low
Analytical sensitivity in plant material (selected matrix)	Concentration (absolute, if possible; relative; low/medium/ high)	*Low dilution of plant material*	High	Medium
Diagnostic sensitivity (comparison of different tests)	%	*Within the same method: More sensitive test*	High	Medium
Analytical specificity	Level	*Specificity: TSWV only*	High	High
(a) Exclusivity (non-target organism): False positives	Level	*0%*	High	High
(b) Inclusivity (target organisms): False negatives	Level	*0%*	High	High
Selectivity	Presence of cross-reactions with matrix	*No*	High	High
Repeatability (near LOD)	Level	*100% at LOD*	High	Medium
Reproducibility/ robustness	%	*100% at LOD*	High	Medium
Applicability				
Applicability in different matrices	Level		Medium	Medium
Amount of material which is included in one sample	Amount of plant units tested		Medium	Medium
Standardised preparation of the reaction (e.g., ready to use reagents)	Yes/no		Low	High
Availability and relevance of controls (in the case of kits)	Yes/no		Medium	High
Protocols				
Available detailed protocols	Yes/no	*Yes*	High	High
Simple test procedure	Yes/no	*Yes*	Low	High
Simplicity of data analysis	Yes/no	*Yes*	Low	High
User friendly test	Yes/no	*Yes*	Low	High
Time needed to complete analysis (<1 h/ 1 day/ several days)	Duration in time unit	*Fastest*	Low	High
Easy to multiplex	Yes/no	*NA*	Low	Low
Database/library dependent (yes/no) (e.g., fatty acids profiling, sequencing)	NA	*NA*	NA	NA

(continued)

Table 3.2 (continued)

Criterion	Descriptor	Target	Relative weight	
			Laboratory	On-site
Chemicals				
Stability of chemicals at ambient temperature	Yes/no	*Yes*	*Low*	*High*
Equipment				
No equipment/ instruments needed (relevant only for on-site tests)	Yes/no	*Yes*	*NA*	*NA*
Test not exclusively developed for a specific instrument	Yes/no	*Yes*	*High*	*High*
Cost of obligatory equipment/ instruments (< €10,000/ €10,000–€50,000/ >€50,000)	Cost (€)	*NA*	*Low*	*High*

LOD, limit of detection

[a]*italics* text, need to be determined by the TPS organiser

here, part of the TSWV genome or an appropriate protein); thus, the choice of a test with the 'wrong' target would have significant impact on the performance. The availability of validation data was considered as moderately important, as the goal was to provide validation data. Analytical and diagnostic sensitivity were not the most important criteria in this case, as the goal was detection and identification in symptomatic plants; i.e., high concentrations of the pest were expected. However, analytical specificity was very important, as it is common that tests for detection and identification of TSWV cross-react with similar orthotospovirus species, and some of these can infect the same plant species (i.e., tomato in this case). On the other hand, as TSWV has a worldwide geographical distribution, it was important that the tests detected all isolates of the virus. As the plan was also to include a considerable number of laboratories for the evaluation of the tests (to subsequently provide enough data for reliable statistical analysis), the requirement for any specific equipment had to be avoided, as this would have led to the elimination of a considerable number of potential participating laboratories. For on-site detection tests, it was important that the reagents needed would be stable at room temperature for a reasonable period of time, and that the test did not require expensive equipment.

As mentioned before, these criteria can also be used in other studies, and to make them easily available to colleagues who might be interested, and an empty form of Table 3.2 can be downloaded with this book (Table 3.1).

3.2.3 Collection of Available Validation Data

After these 'rules' for the selection of the tests have been decided upon, the validation data should be collected to support the selection of the tests. Available validation data and other data about tests (e.g., matrices tested, processing of samples) can be collected through literature searches, internet searches, database searches, experience of the TPS organiser, EUPHRESCO final reports (https://www.euphresco.net/projects/portfolio), dedicated questionnaires or surveys on diagnostic tests used in different laboratories, and validation data from discussions with commercial kit providers. The authors of tests from scientific publications can also be contacted to obtain additional validation data on published tests. Sometimes, validation data for a test will exist, but might not be publicly available, or only be partially available. This communication is time consuming, but it enables a TPS organiser to make the best decisions in the test selection process.

In many cases the validation data for tests are not comparable (e.g., different sample types, units, volumes), and sometimes crucial information is missing or is not available (e.g., sample preparation and concentration). It is important to also evaluate the reported information about the number of tested targets and non-targets and the controls performed, and to compare these between the different tests available. In all of these cases, the expertise of the TPS organiser is invaluable to be able to judge the results reported and the other relevant information.

The biggest designated database of validation data for plant pests is the EPPO database on diagnostic expertise (available at: https://dc.eppo.int/), which contains a significant number of validation datasheets for pests that are considered important in the EPPO region. Laboratories are invited to submit their validation data to this database to help other laboratories when selecting tests, and to contribute to the better diagnostics of certain pests. In addition, the availability of the data can prevent duplication of work across different laboratories, so other laboratories can focus on the pests or tests for which there are no validation data available. Other more specialised databases where validation data can be found are the database of the ISHI-VEG validation reports of the International Seed Federation (available at: https://www.worldseed.org/our-work/phytosanitary-matters/seed-health/ishi-veg-validation-reports/) and the database of the International Seed Testing Association (ISTA) (available at: https://www.seedtest.org/en/method-validation-reports-_con tent%2D%2D-1%2D%2D3459%2D%2D467.html).

In 2021, new sources of validation data for diagnostic kits manufactured by small and medium-sized enterprises (SMEs) became available through the European Plant Diagnostic Industry Association (EPDIA) website. The role of the EPDIA is to ensure the marketability of SMEs by facilitating dialogue with stakeholders and decision makers. In parallel with the establishment of the EPDIA, an EU Plant Health Diagnostics Charter was developed to describe the quality procedures for the production and validation of commercial tests produced by EU manufacturers. This EU Charter will help to guarantee the quality and reliability of products for end users worldwide. The accession of manufacturers to the EPDIA and the EU Charter will allow SMEs to increase their competitiveness (Trontin et al. 2021; EPDIA Quality Charter available at https://www.epdia.eu). The EPDIA database can be searched according to different parameters, such as pests, methods and tests. The EPDIA database contains basic information on the kits (tests), and also validation data sheets provided by the manufacturers and validation data sheets about the availability of their kits in the EPPO database for diagnostic expertise. The database of the EPDIA is available at: https://pestdiagnosticdatabase.eu/.

As mentioned above, for some pests there are a huge number of tests available, including those from commercial providers and from scientific publications. Therefore, the collection of validation data is a time-consuming process that should be planned in advance. For example, for the TPS that was organised for the detection and identification of TSWV, collection of the available validation data took 3 months. TSWV is in the second place on the list of the top 10 economically most important plant viruses (Scholthof et al. 2011; Rybicki 2015), and has a wide host range of >1000 plant species (which include some important vegetables, such as tomato and pepper, and a variety of ornamental plants). This virus has been in the spotlight of diagnosticians and researchers worldwide. Therefore, it was expected that there would be numerous tests available for its detection and identification. After a thorough search through all of the available resources, a total of 76 tests was found, and all available validation data for these were collected. This process included searching the websites of commercial providers, communicating with them to clarify missing data, analysing the available validation data from databases, and collection of the available research articles in which detection of TSWV was included. On the other hand, for some pests that are only important in certain areas, or that are emerging, the relative lack of available tests means that the same step will take considerably less time. As an example, for a TPS on *Cryphonectria parasitica*, the number of tests available including commercial ones and tests from scientific publications was low (only three tests were available at the time).

3.3 Preliminary Studies

Preliminary studies (sometimes called pilot studies) allow TPS organisers to foresee unexpected events and to determine whether the TPS itself is feasible. Preliminary studies are usually carried out initially on a small number of samples. If the results of the small-scale preliminary studies do not meet expectations, or if they show that the study itself is not feasible, the TPS organiser should make appropriate adjustments to prevent further waste of time and resources. For the TPS on detection and identification of TSWV, the study included an evaluation of the possibility to use the same extraction buffer for all ELISA tests, which in all cases differ from the recommended manufacturer's extraction buffers (see Sect. 3.3.3). However, for some kits the results of the preliminary studies showed that this change can affect the performance of the tests. Therefore, this change was not introduced in the TPS. Due to limited resources, the limited number of samples that can be included in a TPS and the tight time schedule, some performance criteria can only be determined in preliminary studies (e.g., analytical specificity). The first step of a preliminary study is usually the definition of the list of tests to be evaluated, with collection of the material to be used for the evaluation of the selected tests (e.g., isolates, strains, populations, plant materials). Then, an assessment of the results of these preliminary studies in comparison to previously adopted criteria is made. The results from preliminary studies can help the TPS organiser to select the best tests to be included in the TPS based on its scope.

3.3.1 Definition of a List of Tests Subjected to Preliminary Studies

There are numerous sources available that can be used to collect tests for any particular pest (see Sect. 3.2.3). Data from all of the available tests should be collected systematically according to the criteria presented in Table 3.2, and their performance should be judged based on those predefined weighted criteria. For this step, the experience and critical judgement of the TPS organiser is crucial to define the tests for preliminary studies. Where there are not enough data collected to make an informed decision on test selection, the TPS organiser can use additional available resources to better define the performance criteria of some tests; e.g., for PCR-based tests, *in-silico* analysis can be carried out to check the specificity of the primers and probes, to make an informed decision on test selection. It is important that all of these selection steps are documented, and the TPS organiser should keep records that explain the selection of the tests.

The selection of the tests for the TPS on the detection and identification of TSWV was a difficult process, considering the number of tests available. Taking into account the scope and predefined criteria, the analytical specificity (i.e., exclusivity, inclusivity) was defined as the most important criterion. The reason for this was that it was known that tomato can be affected by a number of other orthotospoviruses besides TSWV, including alstroemeria necrotic streak virus (ANSV), groundnut bud necrosis virus (GBNV), groundnut ringspot tospovirus (GRSV), tomato chlorotic spot tospovirus (TCSV), tomato yellow (fruit) ring virus (TYRV), tomato zonate spot virus (TZSV), tomato necrotic ringspot virus (TNRV), watermelon silver mottle tospovirus (WSMoV) and capsicum chlorosis orthotospovirus (CaCV) (EFSA, 2012). The symptoms caused by those orthotospoviruses, and also by infection with some other viruses, can be similar, and therefore laboratory testing is needed to identify the causal virus species. In addition, even though TSWV is no longer on the list of quarantine pests in the EU (Regulation (EU) 2019/2072 2019), it is still a very important pathogen and had a status of a regulated non-quarantine pest. The importance of TSWV for the production of agricultural plants is highlighted by the significant losses that can occur as a consequence of TSWV infection, combined with its extensive host range. An increasing problem was the emergence of TSWV resistance-breaking isolates (Turina et al. 2012). The resistance-breaking isolates can overcome the Sw-5 resistance gene in tomato, to cause significant losses, and as this gene provides the only commercially available resistance to TSWV in tomato, it is very important to limit or prevent the further spread of these isolates (Aramburu and Martí 2003; Ciuffo et al. 2005; Turina et al. 2012). Therefore, accurate detection and identification of TSWV is an important step to establish effective control strategies. Based on these criteria, the tests suitable for the detection and identification of TSWV in symptomatic leaves of tomato were defined.

As indicated, after an extensive search for TSWV detection by commercial providers of tests and in scientific papers (with information collected from websites and through direct contact), a total of 76 different tests were evaluated for inclusion in the preliminary studies:

- 13 ELISA tests (DAS-, TAS-, B-fast, ELISA with specific single chain antibodies);
- 2 luminex tests;
- 2 tissue-blot immunoassays (TBIAs);
- 2 dot-blot immunoassays (DBIAs);
- 4 on-site detection tests (lateral flow devices [LFDs], rapid immune gold);
- 2 dot-blot hybridisation tests;
- 36 reverse transcription (RT)-PCR or immunocapture (IC) RT-PCR tests;
- 8 real-time RT-PCR tests (SYBR green, TaqMan);
- 4 RT loop-mediated isothermal amplification (LAMP) or IC-RT-LAMP tests;
- 1 RT thermostable helicase-dependent DNA amplification (RT-HAD) test;
- 1 hyperspectral imaging and outlier removal auxiliary classifier generative adversarial nets (OR-AC-GAN) test;
- 1 microarray test.

All 76 of these tests for detection and/or identification of TSWV were based on the biological, serological and molecular properties of the pathogen. Biological tests such as mechanical inoculation of test plants do not allow pathogen identification, and although widely used, serological detection is often hampered by cross-reactions with other similar orthotospovirus species (Hassani-Mehraban et al. 2016). The molecular tests such as RT-PCR and real-time RT-PCR were developed based on amplification of different genomic parts. Molecular tests are usually more sensitive compared to biological and serological tests; however, they can also cross-react with other orthotospovirus species or they do not detect all TSWV isolates. Consequently, the main challenge for the detection and identification of TSWV was the selection of the appropriate method and test. The process and reasoning for the test selection for the TPS on TSWV was the following.

The ELISA method was taken into consideration because it is widely used and 13 tests were evaluated. Some of these tests were excluded because the commercial provider had stopped production of the test or was in the process of changing the antisera, or because the TPS organiser could not obtain the required information on the validation data despite direct communication with the company. ELISA with specific single-chain antibodies was described in one scientific publication. However, this was excluded because the antibodies are not commercially available. In total, five ELISA tests were selected for the preliminary studies (Table 3.3).

Tests based on Luminex technology were not selected for the preliminary studies because this requires specific equipment that is not available for many diagnostic laboratories, and the TPS organiser did not have experience with this method. TBIA

Table 3.3 The tests selected for the preliminary studies for tomato spotted wilt orthotospovirus. Those then selected for the test performance study are shown in bold text

Method	Tests for validation
ELISA	ELISA 1
	ELISA 2
	ELISA 3
	ELISA 4
	ELISA 5
On-site: Lateral flow device	**Lateral flow device 1**
	Lateral flow device 2
Conventional RT-PCR	RNA PCR reaction kit (commercial)
	Hassani-Mehraban et al. (2016) (RT-PCR generic for orthotospoviruses)
	Hassani-Mehraban et al. (2016)
	Mumford et al. (1994)
	Zarzyńska-Nowak et al. (2018)
	Fineti Sialer et al. (2002)
	Vučurović et al. (2012)
	Panno et al. (2012)
Real-time RT-PCR (RT-qPCR)	Boonham et al. (2002)
	Roberts et al. (2000)
	Debreczeni et al. (2011)
	Mortimer-Jones et al. (2009)

and DBIA tests were not selected for the preliminary studies due to the lack of validation data, and because the interpretation of the results in some cases is difficult, as the results can depend on the experience of the person reading them. In addition, the TPS organiser did not have experience with these methods.

For the on-site detection methods, four tests were taken into consideration, and two were selected for the preliminary studies (Table 3.3). These tests were selected because of their practicality for on-site use. Two of these tests were excluded because there was no commercial kit available.

Altogether, 52 molecular tests were considered. The LAMP method was not selected because the protocols were in Chinese and Japanese only. IC-RT-LAMP tests were not selected because these tests required several steps, and in addition, IC-RT-LAMP is not widely used in diagnostic laboratories in the EU (i.e., there were no EU research publications, and this was not included in the EPPO and IPPC diagnostic protocols for TSWV detection). SYBR green real-time RT-PCR, dot blot hybridisation, RT-HAD, OR-AC-GAN and microarrays were not selected because they were not frequently used in diagnostic laboratories, with the consequent lack of validation data. In addition, OR-AC-GAN and microarrays require specific equipment that was not considered as standard laboratory equipment. Among the 34 conventional RT-PCRs considered, eight were selected for the preliminary studies based on the availability of validation data and based on the results of *in-silico* analysis

(Table 3.3). Among these eight selected tests, seven were considered to be TSWV specific, and one was a generic test for orthotospoviruses. The generic test (Hassani-Merhaban et al. 2016) allowed the detection of American clade 1 orthotospoviruses, which includes TSWV. This generic test was selected because it allowed identification of orthotospovirus species by Sanger sequencing of the RT-PCR product. IC-RT-PCR tests were not selected because the performance of these tests requires additional steps compared to conventional RT-PCR. In addition, IC-RT-PCR is not widely used in diagnostic laboratories in the EU (i.e., there were no EU research publications, and this was not included in the EPPO and IPPC diagnostic protocols for TSWV detection). From the five available TaqMan real-time RT-PCRs, four were selected for the preliminary studies based on the availability of validation data and based on the results of *in-silico* analysis (Table 3.3). One commercial TaqMan real-time RT-PCR test was not selected for the preliminary studies because the protocol was available in Russian only.

In the TPS on TSWV, the validation data collected for the tests varied across companies and publications. For some tests, extensive validation data were available, for others, there were little or none. Therefore, comparisons of the different tests based on the available validation data was very difficult.

3.3.2 Collecting Isolates/Strains/Populations and Plant Material

Depending on the scope of a particular TPS and the availability of different materials and quarantine requirements, different types of samples can be used for evaluation of the tests in a TPS. The material can be obtained from various sources, which include international and national collections, collections of the TPS organiser, and through interlaboratory exchanges. It is very important that the material selected covers the variability in the target. It is highly recommended that reference materials are used whenever possible (see Sect. 3.6.2). If no reference material is available, well characterised material prepared and tested in the laboratory of the TPS organiser should be used (i.e., internal reference material). Considering the seasonal nature of plant production and the possible lack of naturally infected plants or material in the country of the TPS organiser, or if fresh material is not available, or if the material is not available in sufficient quantities for all of the participants, it is possible to use spiked material, where the tested analyte is spiked into the healthy matrix defined in the scope of the TPS. Pure cultures and DNA/RNA extracts can also be used. For more details, see Sects. 3.2.1 and 3.6.2.

The TPS organiser should plan in advance the isolates that will be required for the preliminary studies, and later for the TPS, and how they will be provided. It is important to consider whether a Letter of Authorisation or other import permits might be needed, and to consider the relevant quarantine regulations. If the isolates are obtained through collaboration with a colleague, the TPS organiser should consider signing a Material Transfer Agreement, which is a written contract that governs the transfer of tangible research materials between two organisations when the recipient intends to use them for their own research purposes. All isolates collected for inclusion in the TPS should first be evaluated in preliminary studies.

For the study on TSWV, it was important to include different isolates of TSWV, to cover the different populations of the virus as well as the isolates of other orthotospoviruses. These isolates were obtained from the collection of the TPS organiser and from the commercial collection of the Leibniz Institute DSMZ - German Collection of Microorganisms and Cell Cultures (DSMZ). In addition, some isolates were provided by fellow researchers. In total, 11 species of orthotospoviruses were included in the preliminary studies. TSWV was represented with 15 isolates, impatiens necrotic spot virus (INSV) with five isolates, chrysanthemum stem necrosis virus (CSNV), TCSV and TYRV with two isolates, and ANSV, CaCV, GRSV, iris yellow spot virus (IYSV), melon severe mosaic virus (MSMV) and WSMoV with one isolate each (Table 3.4).

3.3.3   Evaluation of the Tests Selected for the Preliminary Studies

The tests selected for the preliminary studies are evaluated internally by the TPS organiser, to provide missing validation data for the selection of the final tests to include in the TPS, and to identify any difficulties for the TPS organisation. The number of samples included in the preliminary studies can be different than those for the TPS, based on resources, availability of isolates/strain/populations, and the diagnostic parameters for which data are required. Therefore, performance characteristics, such as, inclusivity, exclusivity and selectivity, which usually require many samples to be prepared, can be determined in preliminary studies based on a panel of different isolates/strains/populations of the target organism, with non-target pests that might cross-react with the target organism, and if appropriate with healthy plant samples (i.e., matrix controls). Throughout the whole process, the EPPO guidelines should be followed (EPPO PM 7/98(5) 2021a). Trained or experienced personnel should perform the analyses. During the process, the personnel should be supervised and all of the steps should be carried out to avoid cross-contamination. All of the documents describing the selection process in the preliminary studies should be adequately filed, as well as all of the protocols, including any changes made to them and the reasons leading to the decisions on how the test selection or any test modifications were implemented. The documentation should be stored in accordance with the quality assurance system of the TPS organiser, and should provide traceability of all of the steps.

Some modifications to the test protocols can be made, but these should be first examined in the laboratory of the TPS organiser. For example, when several ELISA tests are available from commercial providers, each provider recommends different

Table 3.4 Results of the preliminary studies for the tests to be chosen for the test performance study for tomato spotted wilt orthotospovirus

Virus-isolate	Detail/ dilution (-fold)	ELISA		Lateral flow device		RT-PCR	RT-qPCR		
		1	2	1	2	Hassani-Merhaban et al.	Boonham et al.	Roberts et al.	Mortimer-Jones et al.
TSWV-PV-0204		pos	pos	pos	pos	pos	pos	pos	pos
TSWV-USA	Tomato	pos	pos	pos	pos	pos	pos	pos	pos
TSWV-108-19	Serbia, tomato***	nt	nt	nt	nt	pos	pos	pos	pos
TSWV-109-19	Serbia, tomato***	nt	nt	nt	nt	pos	pos	pos	pos
TSWV-France 77	Chilli pepper***	pos	pos	pos	pos	pos	pos	pos	pos
TSWV-Italy	Pepper, 2011	pos	pos	pos	pos	pos	pos	pos	pos
	Tomato	pos	pos	pos	pos	pos	pos	pos	pos
	Hot pepper	pos	pos	pos	pos	pos	pos	pos	pos
	Pepper, 2015	pos	pos	pos	pos	pos	pos	pos	pos
	Lisianthus	pos	pos	pos	pos	pos	pos	pos	pos
TSWV-NPPO-NL: 21007721	The Netherlands, liguralia	pos	pos	pos	pos	pos	pos	pos	pos
TSWV-PV-0182		pos	pos	pos	pos	pos	pos	pos	pos
	10	pos	pos	pos	pos	pos	pos	pos	pos
	100	pos	pos	pos	pos*	pos	pos	pos	pos
	1,000	pos	pos	neg	neg	pos	pos	pos	pos
	10,000	pos	pos	neg	neg	pos*	pos	pos	pos
	100,000	neg	neg	nt	nt	pos*	pos	pos	pos
	1,000,000	neg	neg	nt	nt	neg	pos	pos	sus
TSWV-PV-0389		pos	pos	pos	pos	nt	nt	nt	nt
	10	pos	pos	pos	pos	pos	pos	pos	pos
	100	pos	pos	pos	pos	pos	pos	pos	pos
	1,000	pos	pos	pos	pos*	pos	pos	pos	pos
	10,000	pos	pos	neg	neg	pos*	pos	pos	pos
	100,000	neg	neg	neg	neg	pos*	pos	pos	pos
	1,000,000	neg	neg	nt	nt	neg	pos	pos	pos
	10,000,000	neg	neg	nt	nt	neg	neg**	neg**	neg**
	100,000,000	neg	neg	nt	nt	neg	neg	neg	neg**
	1,000,000,000	neg	neg	nt	nt	neg	neg	neg	neg
TSWV-PV-1175	100	pos	pos	pos	pos	pos	pos	pos	pos
	1,000	pos	pos	pos	pos	pos	pos	pos	pos
	10,000	pos	pos	pos	neg	pos	pos	pos	pos
	100,000	neg	neg	neg	neg	pos	pos	pos	pos
	1,000,000	neg	neg	neg	neg	pos	pos	pos	pos
TSWV-PV-0393	10	pos	pos	pos	pos	pos	pos	pos	pos
ANSV-PV-1027		pos	pos	pos	pos*	neg	neg	neg	neg
CaCV-PV-0864		neg	neg	neg	neg	neg	neg**	neg	neg
CSNV-PV-0529		neg	pos	neg	pos*	neg	neg	neg	neg
CSNV-PV-1219		neg	neg	neg	neg	neg	neg	neg	neg
GRSV-PV-0205		pos	pos	pos	pos*	neg	neg	neg	neg
INSV-PV-0280		neg	neg	neg	neg	neg	neg	neg	neg
INSV-PV-0281		neg	neg	neg	neg	neg	neg	neg	neg
INSV-PV-0485		neg	neg	neg	neg	neg	neg	neg	neg
INSV-PV-1123		neg	pos*	neg	neg	neg	neg	neg	neg
INSV-PV-1189		neg	pos	neg	neg	neg	neg**	neg	neg
IYSV-PV-0528		neg	neg	neg	neg	neg	neg	neg	neg**
MSMV-VE440		neg	neg	neg	neg	neg	neg**	neg	neg
TCSV-PV-0390		pos	pos	pos	pos	neg	neg**	neg	neg
TCSV-PV-0391		pos	pos	pos	pos*	neg	neg	neg	neg
TYRV-PV-0526		neg	pos	neg	neg	neg	neg	neg	neg
TYRV-PV-0532		neg	pos	neg	pos*	neg	neg	neg	neg
WSMoV-PV-0283		neg	neg	neg	neg	neg	neg	neg	neg

All PV isolates were obtained from the DSMZ collection

pos, positive; neg, negative; nt, not tested; sus, one replicate positive, one with Cq above the cut-off

TSWV, tomato spotted wilt orthotospovirus; ANSV, alstroemeria necrotic streak virus; CaCV, capsicum chlorosis orthotospovirus; CSNV, chrysanthemum stem necrosis virus; GRSV, groundnut ringspot tospovirus; INSV, impatiens necrotic spot virus; MSMV, melon severe mosaic virus; TCSV, tomato chlorotic spot tospovirus; TYRV, tomato yellow (fruit) ring virus; WSMoV, watermelon silver mottle tospovirus

* weak reaction, close to the limit of detection

** Cq above the cut-off value. For all RT-qPCRs, a Cq cut-off of 35 was used

*** resistance breaking strain

Other isolates were provided by: Branka Krstić, University of Belgrade-Faculty of Agriculture (Serbia); Eric Verdin, INRA (France); Carla Oplaat, NVWA (The Netherlands); Hanu Pappu, Washington State University (USA); Laura Tomassoli, CREA (Italy)

sets of buffers. However, as a TPS organiser is limited in terms of resources and time, some modifications (e.g., the use of the same buffers for all of the ELISA tests) can be introduced to facilitate preliminary studies and as a final result to include more tests in the TPS and to allow more validation data to be produced and to standardise the testing conditions. The producers of commercial tests might benefit from such a study, as getting valuable information on how their tests perform under certain conditions can be of common interest; however, it is recommended that the TPS organiser communicates these adaptations to the kit providers. Communication with kit producers is important because kits are optimised with specific chemicals/buffers, and if changes are made this might affect their performance. Results obtained in that way might not reflect the "true" performance of the kit. The TPS organiser can conduct a small comparison study before any decision on the possibility to use different chemicals/buffers is made.

Regarding the TPS on TSWV, among the 19 tests that were included in the preliminary studies, eight tests were selected for the TPS (Table 3.3, bold text): two DAS-ELISA, two tests for on-site detection, one conventional RT-PCR, and three real-time RT-PCRs (RT-qPCR), each based on predefined criteria. The results of the preliminary studies obtained for the tests selected for the TPS are presented in Table 3.4. Although all available serological tests cross-reacted with other orthotospoviruses, the best in terms of analytical sensitivity were selected because of their robustness and because they were widely used in many diagnostic laboratories (e.g., ELISA tests) or their practicality for on-site detection. For the molecular methods, only tests that did not show cross-reactions with other orthotospoviruses were selected for the TPS.

When the final selection of the tests to be included in the TPS has been achieved, and when these include tests from commercial providers, the decision (with justification) of the TPS organiser should be communicated to the company, respecting confidentiality. In addition, if tests from companies are selected, the TPS organiser should inform the company about the extent of the TPS as soon as the information is available (i.e., number of samples, number of participants); the availability of reagents at the time of the TPS might be a limiting factor, and this should be identified and anticipated by the TPS organiser. This will enable the company to produce enough reagents in time for all of the participants. In addition, this can lead to the establishment of good practice in communication with the companies.

3.4   Selection of the TPS Participants

The selection of competent laboratories is critical for a TPS. This section describes the steps required to select the TPS participants, which includes identification of potential participants, establishment of criteria for selecting participants, invitation for participation of potential participants, and establishing a contract with the participants (Fig. 2.2).

3.4.1 Identification of Potential Participants for a TPS

Potential participants in a TPS can be identified in different ways, such as through surveys, the EPPO database on diagnostic expertise, professional networks, previous participation in proficiency tests or TPS, and social media. Ideally, all laboratories including diagnostic laboratories, private laboratories at commercial companies, and laboratories at public institutions, might have the opportunity to express their interest in taking part in a TPS.

3.4.2 Definition of Weighted Criteria for the Selection of the TPS Participants

The criteria for the selection of the TPS participants must be determined in advance, on the basis of the requirements of the TPS (e.g., see Table 3.5). A target value and the relative weighting need to be assigned to each criterion, to give greater importance to the most critical ones. Weighted criteria need to be set to objectively select the participants for a TPS, with an emphasis on availability of the applicants to perform the tests within the required timeframe, their technical expertise in the use of the methods, authorisation to work with the specific pest, the possibility to obtain import documentation in time (e.g., a Letter of Agreement, if needed), and the quality assurance system in place in the participating laboratory. Other criteria that might be considered in the selection are, for example, technical expertise on the pest group, previous participation in proficiency tests or TPS, or known ability to perform all of the methods selected for the TPS. All criteria considered of high importance must be met by the participants, to be sure that they are proficient and can correctly perform the selected tests, to enable the correct analysis and evaluation of the TPS results. Further criteria (weighted as less important) can be used to allow the objective selection of qualified participants in case there are too many laboratories applying to take part in the TPS.

Depending on the scope of the TPS, the expected number of participants, and the target pest of the TPS, different criteria can be used as the most important for selection of the potential participants. VALITEST TPS organisers considered the following criteria as the most important for selection of the participants: appropriate authorisation to work with a particular pest; appropriate equipment and facilities; commitment to perform analyses on time; willingness to implement all of the tests within the method; technical expertise; and existing quality assurance and traceability. However, exceptions can be made if required; e.g., if the TPS is organised for a 'new' pest, as was the case in the TPS for tomato brown rugose fruit virus, laboratories are expected to lack technical expertise with the pest, and therefore this criterion was not decisive in the selection of potential participants. If a TPS for a pest that has a limited area of distribution is being organised, it can be expected that the number of potential participants will be limited, and therefore, the decisions

Table 3.5 Criteria for selection of participants in the test performance study for tomato spotted wilt orthotospovirus

Criteria[a]	Descriptor[a]	Target	Relative weight*[b]
General information			
Time schedule described in the invitation letter compatible with participant's availability, and participant is committed to perform the analyses and report on the results in the time frame defined	Yes/no	Yes	High
Authorised by the national competent authority to work with the specific pest (viable pest/ inactivated pest/ DNA/ RNA will be shipped)	Yes/no	Yes	High
Traceability in place/ quality assurance in place	Yes/no	Yes	High
Possibility to obtain an import document or letter of authority (EU countries) (only necessary when viable pests are sent)	Yes/no	Yes	High
Possibility to obtain an import document or letter of authority (EU countries) within 4 weeks to receive the samples containing the specific pest (only necessary when viable pests are sent)	Yes/no	Yes	High
Technical expertise for the pest group (e.g., routine analyses, method developments, publications, participation in congresses)	Number of years, or validation data submitted to EPPO database, or other publications	>1 year; advantage if validation data submitted/ published	Medium (validation data: An advantage)
Previous participation in proficiency tests or TPS	Yes/no	Yes	Medium
Constraints for delivery	Yes/no (if yes, explain)	No	Medium
Any problems or limitations with delivery on dry ice	Yes/no (if yes, explain)	Preferably no	Medium
	Yes/no	Yes	Medium

(continued)

Table 3.5 (continued)

Criteria[a]	Descriptor[a]	Target	Relative weight*[b]
Ability/ willingness to perform all of the methods described in the invitation letter (note: It is necessary to perform ALL tests for the selected methods)[c]			
Expertise			
ELISA	*Number of years/ samples, or validation data submitted to EPPO database, or other publications*	>1 year or > 30 samples; advantage if validation data submitted/ published	Medium (validation data: An advantage)
(RT-)PCR	*Number of years/ samples, or validation data submitted to EPPO database, or other publications*	>1 year or > 30 samples; advantage if validation data submitted/ published	Medium (validation data: An advantage)
Real-time (RT-)PCR	*Number of years/ samples, or validation data submitted to EPPO database, or other publications*	>1 year or > 30 samples; advantage if validation data submitted/ published	Medium (validation data: An advantage)
Equipment			
Absorbance reader (company/ model of instrument, wavelength of filters)	*Yes/no (if yes, provide details)*	Yes, with appropriate characteristics	High
Thermal cycler/ gel electrophoresis system/ gel imaging system (company/ model of instrument)	*Yes/no (if yes, provide details)*	Yes, with appropriate characteristics	High
Real-time thermal cycler (company/ model of instrument)	*Yes/no (if yes, provide details)*	Yes, with appropriate characteristics	High

[a]*italics* text, corresponds to the TPS information form
[b]Criteria can be weighted differently to allow the selection of the participants for the defined scope
[c]Performing all tests within one method by TPS participant will enable collection of more validation data which with the application of appropriate statistical analyses will let to the increased relevance and confidence of the TPS results

should be made with more relaxed criteria. Future TPS organisers should have all of these aspects in mind during the planning stage of a TPS.

Additionally, information about participants equipment is required, to help the TPS organiser to plan the TPS and to interpret the results obtained.

3.4.3   Invitations and Analysis of the Eligibility of Potential TPS Participants

After the organiser of a TPS has identified the potential participants for the TPS and defined the criteria that should be met by the participants of the TPS, invitation letters for participation in the TPS should be send out. An invitation letter must contain the name of the pest that will be the target of the TPS, with a description of the scope of the TPS, and the details of which methods will be evaluated in the TPS (Appendix 1). It should also inform the potential participants about the timeline and the deadlines. To assess the eligibility of an interested laboratory for participation in the TPS, the invitation letters should be sent along with a TPS participant information form that asks the potential participants about their experience with the diagnostic methods, pest groups and quality assurance, and about their available equipment (Table 3.5). Potential participants should respect the deadlines to provide the requested information. If a potential participant does not return the form with the requested data filled in by the deadline, it can be considered that they are not interested in taking part in the TPS.

Responses from interested laboratories are then evaluated by the TPS organiser using the weighted criteria described above to select the best TPS participants, and their participation is confirmed by sending them an acceptance email. It is difficult when organising a TPS to estimate how many potential participants will be interested in taking part in a TPS. For the TPS organiser this information is important, because it significantly affects the TPS process, especially in terms of the preparation of the materials and to ensure sufficient participation to allow proper statistical analysis of the data. Therefore, we present here the experience gained in the VALITEST Project based on the 12 TPS organised, which shows that a response rate of 20% to 70% can be expected (Table 3.6). It should be taken into consideration that in some cases, not all of the registered laboratories will be able to provide their results (Table 3.6) for different reason, such as inability to obtain import permits or chemicals needed to perform the analyses on time, and in the case of a global crisis (e.g., the COVID-19 outbreak).

Participant selection can significantly affect the final results. It is generally acknowledged that laboratories participating in a TPS should be proficient so that the results obtained reflect the performance of the tests and not the proficiency of laboratories. Thus, the criteria and weighted values used for the selection of participants should be well established to select proficient laboratories. If a participant provides results that are far from the expected results, that particular dataset may also be excluded from the analysis. However, the selected criteria and weighted values should not lead to the exclusion of too many laboratories. Indeed at least 10 datasets per test are needed to perform the statistical analysis and to provide results with high confidence (EPPO PM 7/122(1)). In addition, if the criteria used are too strict, only highly proficient laboratories will be selected and a high proportion of concordant datasets can be expected. As a consequence, the performance characteristics of the

Table 3.6 Response rates for the test performance study targets organised in the VALITEST Project

Pest	TPS organiser	Number of invited participants	Number of participants registered	Number of participants who submitted results	Percentage (%) of registered participants
Bursaphelenchus xylophilus	ANSES[a]	31	21	21	68
Citrus tristeza virus	ANSES[b]	34	17	15	50
Cryphonectria parasitica	UNITO	47	11	10	23
Erwinia amylovora	NIB	76	32	30	42
Fusarium circinatum	FERA	28	20	20	71
Pantoea stewartii subsp. stewartii	NIB	76	23	21	30
Plum pox virus (lab tests)	NVWA	37	17	17	46
Plum pox virus (on-site tests)	ANSES[c]	41	15	14	37
Tomato brown rugose fruit virus	CREA	96	34	31	35
Tomato spotted wilt orthotospovirus	NIB	92	21	19	23
Xanthomonas citri pv. Citri	ANSES[b]	43	19	18	44
Xylophilus ampelinus	FERA	39	12	11	31

ANSES: Plant Health Laboratory, French Agency for Food, Environmental and Occupational Health and Safety, [a]Nematology Unit, Le Rheu, France, [b]Unit for Tropical Pests and Diseases, Saint Pierre, Reunion island, France, [c]Quarantine Unit, Lempdes, France; UNITO: AGROINNOVA - Centre of Competence University of Torino, Grugliasco, Italy; NIB: National Institute of Biology, Ljubljana, Slovenia; FERA: Fera Science Ltd., York Biotech Campus, Sand Hutton, York, United Kingdom; NVWA: Dutch National Plant Protection Organisation, National Reference Centre, Wageningen, The Netherlands; CREA: Consiglio per la Ricerca in Agricoltura e l'Analisi dell'Economia Agraria (CREA), Centro di Ricerca Difesa e Certificazione, Rome, Italy

test(s) might be overestimated and might not reflect the global diagnostic community.

Usually in a TPS this decision is based on the self-assessment of competence of the potential participants, which can be biased and subjective, but this is a risk that is accepted by TPS organisers. It is worth mentioning that during the selection process, it is important to have open communication with the participants, as many misunderstandings can be resolved and competent laboratories will not be excluded; on the other hand, unqualified laboratories will not participate in the TPS. The whole process therefore requires dedicated time that should be appropriately considered during the planning of the TPS.

3.5   Contracts and Technical Information for TPS Participants

A TPS organiser should provide all of the relevant information regarding the TPS to the participants, which includes the definition of the scope of the TPS, the contract and the technical sheets, including information regarding specific requirements, such as a Letter of Authorisation. The TPS organiser is required to treat the information on the participants as confidential, if this status is not defined differently.

A TPS organiser should foresee potential difficulties in the course of the TPS, and therefore should provide the information regarding these to the TPS participants in advance. Among the difficulties that can affect the course of a TPS, the most important and frequent include delays in obtaining the authorisation document (e.g., a Letter of Authorisation) required for quarantine pests and in obtaining the necessary chemicals and reagents. However, if the TPS organiser informs the participants of those potential difficulties sufficiently in advance, their impact can be avoided, or at least minimised.

The contract serves to define the rights and obligations of the parties involved in a TPS, (i.e., the organiser and the participating laboratory). In addition, the contract contains a detailed description of the timelines and the detailed conditions of participation for the interested laboratories. The TPS contract can also serve as a registration form. An example template of a contract is given in Appendix 2. Appendix 3 includes a template that can be used to collect the contact information of participating laboratories which might be used in future steps of a TPS (such as sending of the samples and submission of the reports). There are different ways to collect such data, and in the framework of VALITEST, they were collected using an MS Excel template.

Contracts need to be accompanied with the TPS technical sheet that contains a general overview of the TPS, with the required information about the tests, sample panels and important dates (planning of the TPS), and the detailed experimental protocols for performing each test. The technical data sheet should also include a list of all of the necessary consumables and the quantities that will need to be ordered by the participants of the TPS. An example template of a TPS technical sheet is given in Appendix 4.

The TPS organiser should establish and maintain open and transparent communications with the TPS participating laboratories. Although this can be time consuming for both sides, it can be crucial to avoid misunderstandings.

3.6 Preparation and Dispatch of Samples and Reagents

For TPS (as for any other interlaboratory comparisons), it is important that the samples match as closely as possible the materials encountered in routine testing, if it is not defined differently in the aim and the scope of the TPS. This includes the matrix (e.g., host plant), the target (e.g., pest) and the concentrations (e.g., infection levels) (EPPO PM 7/122(1)).

3.6.1 Definition of the Panel of Samples

Depending on the scope of a particular TPS, different types of samples can be prepared, such as fresh naturally infected plant material, DNA/RNA, spiked matrix, artificially inoculated matrix, freeze dried infested plant material, samples that mimic infested material, pure cultures, traps and pest specimens. It is very important that the panel of samples is appropriate for each diagnostic method included in a TPS, depending on the pest/ matrix combination, covering:

- the range of concentrations and genetic diversity of the target pest;
- the diversity of the uninfested material (when relevant) for the selectivity assessment;
- the diversity of the non-target organisms that occur in the same ecological area.

In addition, the quantities of samples prepared for the TPS needs to be sufficient to enable evaluation of the selected performance characteristics in-house and interlaboratory, and the homogeneity and stability testing (see Sect. 3.6.3). When possible, TPS organisers should prepare additional samples to cover unexpected events that can occur during a TPS, such as damage during transport and prolonged stability testing requirements. For example, additional stability testing was required during the VALITEST Project. Due to the COVID-19 outbreak, some laboratories were not able to perform the analyses in a timely manner. By performing additional stability testing, TPS organisers confirmed that the samples were still fit for purpose at the time when the participating laboratories submitted their results.

The numbers of samples sent to the participating laboratories should be sufficient to allow statistical evaluation of the performance characteristics of interest. However, the participating laboratories should not become overloaded, so the TPS organiser should take into account the time and resources participants need to invest (e.g., having a maximum of 24 samples including the controls, which corresponds to the number of tubes that can be used in a standard laboratory centrifuge).

The samples should be accompanied by detailed instructions on how the participating laboratories should handle them upon receipt. The TPS organiser is responsible for providing appropriate coding of the samples. If possible, sample codes should be different for each of the participating laboratories to avoid disclosure of the identities of the participants.

Ideally, a standardised sample panel should be used when possible. This contributes to a more robust comparison of the data obtained in different validation studies, and to the application of standardised statistical analysis of the data. However, each TPS organiser can create their sample panel based on the performance criteria that should be evaluated during the TPS. Recently, significant effort has been invested in the standardisation of sample panels to ensure that they are suitable for statistical analysis of the data, in addition to calculation of the standard diagnostic parameters, as explained in EPPO PM 7/122(1).

In the framework of VALITEST, a sample panel was proposed (Massart et al. 2022) and listed the following recommendations:

- a minimum of five dilution points from a sample that contains the target pest and three replicates of the serial dilutions for the determination of the analytical sensitivity using the probability of detection model. The range of serial dilutions must cover the limit of detection, as determined during the preliminary studies by the validation organiser.
- a minimum of three samples free from the target pest (i.e., negative samples) and two samples infected with the target pest (i.e., positive samples), which should be independent of each other, and which are used for determination of the diagnostic sensitivity and diagnostic specificity. At least one positive sample should have a low concentration of the target pest; e.g., close to the limit of detection estimated during the preliminary studies by the validation organiser.
- a minimum of two replicates for each of three negative samples and two positive samples, with the samples independent of each other, for determination of the repeatability and reproducibility using the accordance and concordance of Langton et al. (2002). By using accordance and concordance, it is possible to determine if a particular laboratory performs poorly, or if a particular sample or test is performed poorly (see more details in Massart et al. 2022).
- to determine the analytical specificity of a test, it is very important to have closely related species included in the sample panel (at the expense of the non target).

The composition of a sample panel is heavily dependent on the performance criteria that need to be evaluated within the TPS. For example, if a sample panel only contains samples with high concentrations of the target and/or samples with concentrations close to the limit of detection, the test results of the TPS will only provide the reproducibility at the extremes of the target concentrations. Thus, this will not provide an answer to the question of whether the test is appropriate to monitor the infection status in particular host species. If the aim is to provide data on the analytical specificity of the test and cross-reactions with closely related organisms are expected, these need to be included in the sample panel. In addition, the sample panel needs to contain appropriate controls to monitor the processes (in the laboratories of the TPS organiser and the TPS participants).

Prior to sending the samples to the participants, the TPS organiser needs to define or establish the assigned values for the samples, i.e., the value attributed to a particular property of an interlaboratory test sample (EPPO PM 7/122(1)). The

TPS organiser is also responsible for documenting the procedure in which the assigned values were determined.

Assigned values can be given to the test items in various ways, with the two most commonly used being to assign the reference values on the true health status of the test items, or to assign values based on the results of the tests in preliminary studies which are also expected to be used by the participants later in the TPS. In some cases, when samples are between the positive and negative thresholds of a test or the specimens show overlapping morphological characters, the assigned value can be declared as 'inconclusive' (EPPO PM 7/122(1)).

For the TPS for the detection and identification of TSWV, different strategies were used for the different methods. For the molecular tests, the samples were defined as positive when they contained TSWV, and negative if they contained other orthotospoviruses or if they were without any virus. However, for ELISA and the LFD tests, high dilutions of samples containing TSWV were considered as negative if the best performing test gave negative results under extensive testing during the preliminary studies.

In the TPS for TSWV, the full panel included 22 samples (S-1 to S-22), one or two positive controls (i.e., positive isolation/ amplification controls), and one negative control (i.e., negative isolation control) (Table 3.7). The sample panel was assembled with some slight modifications from the proposed sample panel described above to obtain data on analytical specificity (i.e., five isolates of similar orthotospoviruses were included at the expense of the negative samples and the heavily contaminated positive samples). The expected results were determined based on the results of the preliminary studies.

3.6.2   Reference Material

To prepare the samples of a TPS, the use of reference material is recommended. In the field of plant health, the commercial availability of reference material or certified reference material is limited, and consequently reference material might need to be produced by individual diagnostic laboratories, or by companies that offer positive or negative controls as part of their kits (EPPO PM 7/147 2021b). A reference material in the sense of the ISO is any material that is sufficiently homogeneous and stable with respect to one or more specified properties, and where the suitability for its intended use has been established in a measurement procedure (ISO Guide 30:2015). The term "reference material" is a generic term, where the properties might be quantitative or qualitative. The uses here can include calibration of a measurement system, assessment of a measurement procedure, assignment of values to other materials, and quality control. Accordingly, a certified reference material is defined as a material characterised by a metrologically valid procedure for one or more specified properties and accompanied by a certificate that includes the value of the specified property, its associated uncertainty, and a statement of metrological traceability (ISO Guide 17034). The International Vocabulary of Metrology gives

Table 3.7 Composition of test panel provided to the participants of the test performance study for tomato spotted wilt orthotospovirus

Type of sample	Sample (DSMZ code)	Dilution (-fold)	Sample designation					Health status of TSWV	Assigned values for TSWV		
			ELISA		Lateral flow device		Molecular		ELISA	Lateral flow device	Molecular
			1	2	1	2					
	Healthy tomato 1		S-4	S-8	S-15	S-4	S-7	neg	neg	neg	neg
Healthy tomato	Healthy tomato 1		S-15	S-10	S-20	S-10	S-17	neg	neg	neg	neg
	Healthy tomato 2		S-6	S-9	S-10	S-3	S-21	neg	neg	neg	neg
	ANSV (PV-1027)		S-19	S-14	S-11	S-7	S-12	neg	neg	neg	neg
Other orthotospoviruses	CSNV2 (PV-0529)		S-7	S-19	S-3	S-21	S-15	neg	neg	neg	neg
	GRSV (PV-0205)		S-16	S-11	S-13	S-14	S-2	neg	neg	neg	neg
	INSV2 (PV-0281)		S-9	S-3	S-7	S-2	S-13	neg	neg	neg	neg
	TCSV (PV-0390)		S-12	S-7	S-21	S-15	S-10	neg	neg	neg	neg
	TSWV (PV-1175)	1,000,000	S-14	S-4	S-2	S-5	S-18	pos	neg	neg	pos
	TSWV (PV-1175)	1,000,000	S-20	S-12	S-18	S-22	S-20	pos	neg	neg	pos
	TSWV (PV-1175)	100,000	S-10	S-6	S-17	S-6	S-1	pos	neg	neg	pos
	TSWV (PV-1175)	100,000	S-11	S-21	S-22	S-9	S-19	pos	neg	neg	pos
TSWV dilution series	TSWV (PV-1175)	10,000	S-13	S-2	S-8	S-13	S-11	pos	pos	pos	pos
	TSWV (PV-1175)	10,000	S-21	S-22	S-9	S-19	S-14	pos	pos	pos	pos
	TSWV (PV-1175)	1,000	S-18	S-13	S-1	S-1	S-3	pos	pos	pos	pos
	TSWV (PV-1175)	1,000	S-22	S-15	S-16	S-16	S-9	pos	pos	pos	pos
	TSWV (PV-1175)	100	S-5	S-1	S-4	S-12	S-6	pos	pos	pos	pos
	TSWV (PV-1175)	100	S-8	S-18	S-6	S-20	S-22	pos	pos	pos	pos
	TSWV (PV-0182)	1,000	nt	nt	nt	nt	S-8	pos	nt	nt	pos
TSWV medium concentration	TSWV (PV-0182)	1,000	nt	nt	nt	nt	S-4	pos	nt	nt	pos
	TSWV (PV-0182)	100	S-1	S-5	nt	nt	nt	pos	pos	nt	nt
	TSWV (PV-0182)	100	S-2	S-17	nt	nt	nt	pos	pos	nt	nt
	TSWV (PV-0182)	10	nt	nt	S-14	S-8	nt	pos	nt	pos	nt
	TSWV (PV-0182)	10	nt	nt	S-19	S-17	nt	pos	nt	pos	nt
	TSWV (PV-0389)	100,000	nt	nt	nt	nt	S-5	pos	nt	nt	pos
TSWV at concentrations close to the limit of detection	TSWV (PV-0389)	100,000	nt	nt	nt	nt	S-16	pos	nt	nt	pos
	TSWV (PV-0389)	10,000	S-3	S-16	nt	nt	nt	pos	pos	nt	nt
	TSWV (PV-0389)	10,000	S-17	S-20	nt	nt	nt	pos	pos	nt	nt
	TSWV (PV-0389)	1,000	nt	nt	S-5	S-11	nt	pos	nt	pos	nt
	TSWV (PV-0389)	1,000	nt	nt	S-12	S-18	nt	pos	nt	pos	nt
	Healthy tomato 2		NC	NIC	NC	NC	NC/NIC	neg	neg	neg	neg
	TSWV (PV-0393)	10	PC	PIC	PC	PC	PC/PIC	pos	pos	pos	pos
Controls	No template control*		nt	nt	nt	nt	NAC	neg	nt	nt	neg
	Total RNA (PV-0182; PV-0389)		nt	nt	nt	nt	PAC	pos	nt	nt	pos

nt, not included in the sample panel

TSWV, tomato spotted wilt orthotospovirus; ANSV, alstroemeria necrotic streak virus; CSNV, chrysanthemum stem necrosis virus; GRSV, groundnut ringspot tospovirus; INSV, impatiens necrotic spot virus; TCSV, tomato chlorotic spot tospovirus

PC, positive control; NC, negative control; NIC, negative isolation control; PIC, positive isolation control; PAC, positive amplification control; NAC, negative amplification control; IC, internal control

[a]not provided for participants

the following definition of a reference material: "a material, sufficiently homogeneous and stable with reference to specified properties, which has been established to be fit for its intended use in measurement or in examination of nominal properties" (Anonymous, 2008). Reference materials provide essential traceability in testing, and they are used, for example: (i) for detection and identification; (ii) to demonstrate the accuracy of the results; (iii) to calibrate or verify equipment; (iv) to monitor laboratory performance; (v) to validate or verify tests; and (vi) to enable comparisons of tests (EPPO PM 7/84(2) 2018b).

In the VALITEST Project, a list was drafted of the criteria to be considered for the production of reference materials to be used in TPS (Chappé et al. 2020; Trontin et al. 2021). If needed, TPS organisers can include additional criteria in the list based on their own experience. The identified criteria are the intended use of the material, and its identity, commutability level (i.e., level of agreement between test results obtained with a biological reference material and with an authentic sample), traceability (i.e., information on the source of the material), homogeneity (e.g., non-homogenous material might introduce measurement uncertainties), stability (which should be determined with the 'worst-case scenario' approach; e.g., using the PCR test that targets the longest DNA fragments), assigned value (i.e., expected result of the test) and purity (which describes the presence or absence in the biological reference material of components that might interfere with the results of a test, including non-target organisms, and when relevant, components of the matrix; e.g., false positive results) (Chappé et al. 2020; EPPO PM 7/147 2021b). Depending on the intended use, the reference material might need to fulfil all of the criteria, or some criteria might not be relevant (Chappé et al. 2020; EPPO PM 7/147 2021b).

To describe the reference material used in the TPS for detection and identification of TSWV, a set of descriptors were adapted from Chappé et al. (2020), as shown in Table 3.8.

3.6.3   Stability and Homogeneity Studies

To ensure the stability and homogeneity of the samples included in a TPS, the TPS organiser should test the complete batch of samples. If this is not feasible, representative samples should be selected, and the selection should be justified and documented. If the TPS organiser is providing the chemicals (e.g., primers/ primers-probe separately or in mixes), their homogeneity and stability should also be tested. Sample homogeneity should be tested after the samples are fully prepared and ready for distribution to the participants, but before they are shipped (EPPO PM 7/122(1)). If this is not feasible to include a complete batch of the samples in homogeneity testing, according to currently available recommendations at least 10 randomly selected samples are tested in duplicate (for each pest/matrix/infestation level, including negative samples) (see also ISO 13528).

According to EPPO PM 7/122(1), the TPS organiser should also demonstrate that the samples are sufficiently stable to ensure that they do not undergo significant changes throughout the TPS, including during their storage and transport. If necessary, and especially if the transport requires special conditions (e.g., dry ice), stability tests should be performed under conditions that mimic the transport and storage conditions. Alternatively, the samples can be sent to the participant with the

Table 3.8 Descriptors and their corresponding values for the reference materials prepared for the test performance study for tomato spotted wilt orthotospovirus

Descriptor	Value	Minimum criterion	Description
Intended use	Should be defined (where it is the same as the preparation of reference materials for the scope of the individual tests or the test performance study)	Yes	Reference material for the test performance study on detection and identification of tomato spotted wilt orthotospovirus in symptomatic tomato leaf material
Identity	Identified to the level of internationally recognised diagnostic protocols (mention tests and outcomes)	Yes	All isolates used for the test performance study were obtained from German collection of microorganisms and cell cultures GmbH (DSMZ). DSMZ carried out the isolate characterisation. However, the identities of the virus isolates used in the test performance study were confirmed also in the laboratory ot the test performance study organiser by sanger sequencing of PCR products and by all of the tests included in the test performance study
Traceability	Traceability to a specimen from a reference culture collection	No	Yes. DSMZ collection: Tomato spotted wilt orthotospovirus isolates: PV-0182, PV-0389, PV-1175, PV0393; other orthotospovirus isolates: PV-0390, PV-0529, PV-1027, PV-0281, PV-0205
	Traceability to a specimen from a working culture collection	No	/
	Traceability provided for the target pest and matrix used (the latter, if relevant)	Yes	Samples were prepared by mixing virus isolates (see above) with healthy tomato cv. Money maker grown under greenhouse conditions. The status of these tomato plants was confirmed by all of the tests included in the test performance study
Commutability level	Naturally infested plant material	No	/
	Artificially infested plant material	No	✓
	Spiked plant material	No	✓
	Purified organisms	No	/

(continued)

Table 3.8 (continued)

Descriptor	Value	Minimum criterion	Description
	Total nucleic acids from a sample (target organism in background)	No	/
	Purified nucleic acids	No	/
	Synthetic nucleic acids	Yes	/
Homogeneity	Homogenous; provide tests and test results	Yes	Plant materials were homogenised and several aliquots were prepared. The homogeneity of each batch was confirmed by testing some randomly chosen aliquots with all of the tests included in the test performance study
Stability	Stable	Yes	Stability testing was conducted at several times (for details see Sect. 3.5.3) under conditions that mimic transport and storage conditions. This was done on randomly chosen aliquots of each batch of the samples
	Stability - short term	No	Done when the first participant did the analysis
	Stability - long term	No	Tested on the deadline to perform the analysis. Due to the COVID-19 situation, the stability testing was done at several times also after the deadline to perform the analysis (to 1 July, 2020)
Assigned value	Absolute concentration known	No	No
	Level of concentration known (high/medium/low)	No	Estimated by real-time PCR analysis
	Qualitative status known (above limit of detection)	No	The minimum level of detection that guaranteed all of the repetitions were positive
	Originating from plants with known health status with a recent test result (a given period of time depends on the plant-pest combination and previous experience)	Yes	See above (all assigned values were determined by DSMZ and confirmed by the test performance study organiser)
Purity	Absence of non-targets	No	/
	Absence of interfering non-targets	No	/
		No	/

(continued)

Table 3.8 (continued)

Descriptor	Value	Minimum criterion	Description
	Known ratio of target *versus* non-target interfering with the test - high		
	Known ratio of target *versus* non-target interfering with the test - medium	No	/
	Known ratio of target *versus* non-target interfering with the test - low	Yes	Purities of the isolates were tested by DSMZ. No additional tests regarding the purities of the isolates were carried out by the test performance study organiser (only sanger sequencing of orthotospovirus specific PCR product was done).

✓, used in the test performance study / -, not used in the test performance study

adverse environmental or transport conditions, and then returned unopened for testing to the TPS organiser laboratory. Stability should be tested before samples are dispatched (to ensure that the samples are stable enough to be included in the TPS). Generally, stability testing before sample dispatch is coupled with the homogeneity study, to spare resources and avoid doubling the work in the laboratory. In some cases (e.g., when the samples are prepared as an extract of the pest), based on the previous experience with the pest and taking into account the timeframe of the TPS, when the TPS organiser considers that the stability of the samples can be affected, stability testing can be performed also during the various stages of the TPS. This practice retains as many as possible of the datasets for the final evaluation. During the TPS on detection and identification of TSWV (where the samples were prepared as extracts), stability tests were performed each week starting 7 days after the samples were dispatched to the participants. Stability testing should also be conducted after the deadline for participants to perform the analysis, to confirm that the stability of the samples has been maintained throughout the TPS. In some cases, the TPS organiser has to be prepared to extend the stability studies, because some TPS participants might not manage to perform the analysis before the specified deadline (e.g., as in the case of VALITEST, there was a global pandemic that affected the normal workflow in the majority of the laboratories involved). For material that have been shown to be stable over time (e.g., *Globodera* spp. cysts, some fungal spores), stability testing is not required. If it is not possible to test whole batches of samples, currently available guidelines recommend, when possible, to test a minimum of three randomly chosen samples in duplicate (for each pest/matrix/infestation level, including the negative controls) (see also ISO 13528).

The number of samples tested for the stability and homogeneity might deviate from the recommendations, and this should be documented. If samples are not stable and homogeneous, they can affect the assigned values, and their impact on the evaluation of the results should be estimated. Such samples might be rejected for use in the TPS, or they can be used under specific conditions that should be clearly stated. If the participating laboratories evaluate measurements or tests on test items (e.g., the samples) that are not considered stable any more, these datasets should be excluded from further evaluation (e.g., if the laboratory returns the results too late).

For the TPS for TSWV, stability was carefully assessed because a plant extract known to be unstable to orthotospoviruses was used. During the preliminary studies, aliquots of the last three positive dilutions of one TSWV sample were stored at -20 °C and analysed with the eight tests after 2, 5 and 8 weeks of storage. The results were similar to those obtained with freshly prepared extracts. Stability was also evaluated under conditions that mimicked the transport and storage conditions. The samples were stored at different temperatures and for different times before being tested: for up to 18 weeks at less than -15 °C; and for 2 weeks at less than -15 °C, 3 days on dry ice, and then 5 weeks at less than -15 °C. For the stability and homogeneity testing prior to shipment of the samples and throughout the period of the TPS, six randomly selected aliquots of all of the samples and primer/primer-probe mixtures prepared for the TPS were selected and analysed using the tests included in the TPS. The last stability testing was conducted on the deadline to perform the analysis, which was prolonged due to the COVID-19 situation. All of the results were documented and stored.

3.6.4 Dispatch of Samples and Reagents

The TPS samples need to be dispatched together with the instruction sheet, and along with an acknowledgement of receipt of the panel of samples (for immediate return), and a results form for the participants to fill in after performing the tests. The TPS instruction sheet is intended to help the TPS participants upon their receipt of the parcel, for the identification of the samples, for their storage and analysis, for any special precautions needed, and for the submission of the results.

The following information should be included in the instruction sheet:

- Requirements on how to handle the test items, controls, and/or reagents in the participant laboratory (storage upon receipt);
- Details about the number of sample panels and test items received by the participating laboratory;
- Information about the test items and samples, and for which test they should be used;
- Panel code (participant ID);
- Explanation of how the samples were coded for the different tests or methods;
- The time schedule of the testing;
- The detailed instructions to prepare and condition the test items;

- Other instructions if needed, such as safety requirements, special precautions for handling and destroying of material;
- Specific and detailed instructions for entering, recording and submitting the results and the associated deviations and difficulties;
- The latest date when the results should be submitted;
- The TPS organiser contact information;
- The time frame for the release of the TPS report to the participants.

An example template of an instruction sheet is given in Appendix 5.

Another document that should be given to the participants along with the dispatch of the samples is the acknowledgement of receipt of the panel of samples. With this document, the TPS organiser wants to determine whether the transport conditions might have affected the test items. An example of an acknowledgement form developed by the partners of the VALITEST Project and used in the TPS for the detection and identification of TSWV is given in Appendix 6.

To minimise the possibility of human error in processing the TPS results, the TPS organiser needs to prepare a results form that can be easily manipulated and used to extract the data provided by participants (some examples of results forms are given in Appendix 7). The appropriate format for this document can be MS Excel, using drop-down menus to enter the data where possible, and formulas to calculate the means where appropriate. Also, online recording of the results on different platforms (e.g., google sheets or similar) can be used especially in the cases with a high number of tests and/or participants involved, to enable easier collection, processing and interpretation of the results. In addition to the results of the sample analysis, the participants should be asked to provide other data that are needed to contextualise the results provided, and to interpret the results appropriately.

The TPS organiser should take into account the different policies and laws in the countries of the TPS participants regarding the shipment and receipt of plant materials, especially packages that need to be provided with import permits (e.g., Letters of Authority) due to the presence of quarantine pests in the samples. Where the TPS test items need to be sent on dry ice, it is very important for the TPS organiser to use the services of a reliable courier, to ensure that the samples are delivered within an appropriate time frame and without damage. Some countries have limitations on the acceptance of packages on dry-ice, and some have long customs procedures that can affect the stability of the samples. Therefore, the TPS organiser should anticipate such problems in advance, and appropriate measures should be taken to minimise any impact or damage that would require exclusion of some datasets from the TPS.

3.7 Collecting the TPS Results and Analysing the Data

The TPS participants should send back the completed results forms to the organiser or complete an online results form. The TPS organiser analyses the TPS results and summarises the conclusions on the performances of the tests. During the analysis of the results, the TPS organiser might have to exclude all or some results from some TPS participants; e.g., if there is suspicion of contamination during the processing of the samples, or if the analysis was not performed according to the protocols provided. If there were any deviations from the recommended protocols by any of the participants, this should be recorded, and their potential impact on the results should be assessed.

The performance of the tests is evaluated through determination of their performance characteristics, and through comparisons of the performance characteristics of the different tests (or methods, if the same panel of samples was used). To perform the correct statistical analysis, a TPS requires a minimum number of participating laboratories; i.e., at least 10 valid laboratory datasets per test (EPPO PM 7/122(1); Chabirand et al. 2017). However, it is recognised that this might be a constraint in plant-pest diagnostics. If the results of less than 10 laboratories were used to evaluate the performance of the tests, the report should contain a disclaimer that the results might not be reliable, and that any deviation might be the result of the processing of an insufficient number of datasets.

The main performance characteristics for validation of tests include the following (EPPO PM 7/98(5) and PM 7/122(1)):

- Analytical sensitivity
- Analytical specificity, inclusivity
- Analytical specificity, exclusivity
- Diagnostic specificity
- Diagnostic sensitivity
- Selectivity
- Repeatability
- Reproducibility
- Robustness
- Accuracy

After obtaining the data from the TPS participants, the TPS organiser needs to carry out the full evaluation. The first step is the definition of the outlier results. The TPS organiser can decide how to define the outliers on a case-by-case basis. Frequently, controls provided to participants with the test panel are used as a quality check of the datasets, and only the datasets with concordant results for all the controls are considered as valid. Additionally, the TPS organiser can use other methods to identify outliers; e.g., for the TPS for TSWV, the results of the healthy tomato samples were assessed, and datasets of laboratories where there were two or more (out of three) false positive results for the healthy tomato samples were also excluded. Another option to identify outliers can be to use different graphical

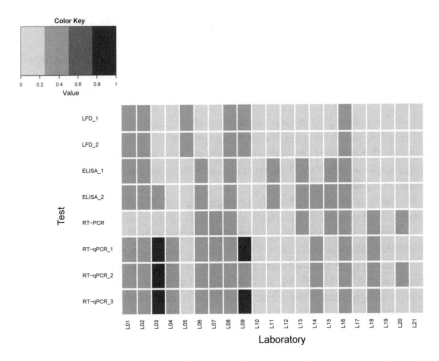

Fig. 3.2 Identification of outlier results from non-target samples. Grey squares, not tested. The heatmaps colour pallet for the cells ranges from light orange (all concordant results) to dark orange (all non-concordant results)

tools, such as heat maps, which allow rapid visualisation of outlier results that are too far away from the results of the other laboratories (Fig. 3.2). If needed, in particular cases, additional outliers can be defined, such as incomplete datasets. Incomplete datasets and datasets with incorrect results of controls can be considered as the first level of outliers to be excluded. Outliers identified only through graphical analysis should be checked carefully in terms of whether they should indeed be excluded or whether they are a part of the method robustness.

The next step in the results analysis is to evaluate the performance of the individual tests. First, the performance can be described in terms of the number and proportion (%) of the results that were inconclusive, true negatives (negative agreement), false negatives (negatives deviations), true positives (positive agreements) and false positives (positives deviations). The number and proportion (%) of concordant and non-concordant results can also be calculated. Inconclusive results can be treated as non-concordant. When it is not possible to reasonably interpret inconclusive results, these can be excluded for the calculation of the performance characteristics.

The calculations can be carried out according to the layout of Table 3.9.

The following parameters can be calculated:

Table 3.9 Layout for determination of the performance characteristics

		True value	
		Positive	Negative
Determined value	**Positive**	True positive	False positive
	Negative	False negative	True negative

- Diagnostic sensitivity = true positives / (false negatives + true positives);
- Diagnostic specificity = true negatives / (false positives + true negatives);
- False positive rate = false positives / (false positives + true negatives) = 1 - diagnostic specificity;
- False negative rate = false negatives / (false negatives + true positives) = 1 - diagnostic sensitivity;
- Relative accuracy = (true positives + true negatives) / total number of samples;
- Power = true positives / assigned positives;
- Positive predictive value = probability that subjects with a positive screening test actually have the disease = true positives / (true positives + false positives);
- Negative predictive value = probability that subjects with a negative screening test actually do not have the disease = true negatives / (false negatives + true negatives);
- Diagnostic odds ratio, as a measure of the effectiveness of a diagnostic test. This is defined as the ratio of the odds of the test being positive if the subject has a disease, relative to the odds of the test being positive if the subject does not have the disease. It is thus calculated as: Diagnostic odds ratio = (true positives / false positives) / (false negatives / true negatives).

Harmonising of the analysis and presentation of the results of a TPS can enable easier analyses of different TPS reports. In the framework of VALITEST, significant efforts were made to find the appropriate way to harmonise the presentation of the results from different TPS. This will enable future TPS organisers to present the main results of their TPS in the same (or similar) way(s).

As each TPS has its own specificities, problems and unexpected obstacles, the TPS organiser should apply appropriate measures to limit or prevent these from having any impact on the final result of the TPS. For example, in the TPS for TSWV, there were problems during the analysis of the raw data obtained from the TPS participants for the real time RT-PCR tests. It was apparent by the "raw" results received from the TPS participants that some results were not interpreted correctly (taking into account all provided Cq values), and some laboratories had clear problems with contamination. Therefore, some of the results submitted by TPS participants were corrected to retain as many datasets as possible in the analysis. This was necessary because the TPS participants used different approaches to set up Cq cut-off values: some set the Cq cut-off value at a too low level, while some did not use any Cq cut-off value even though they had some late signals in the negative controls. Cq cut-off values are equipment, material and chemistry dependent, and need to be verified in each laboratory before the tests are implemented. However, it

Table 3.10 Example of the analysis of the results of a single test (results of RT-PCR are shown)

Test items	Sample	DSMZ code	Dilution (-fold)	INC	TN	FP	FN	TP	INC %	TN %	FP %	FN %	TP %	Concordant	Non-concordant	Concordant %	Non-concordant %
S-7	Healthy tomato 1			1	12	0			7.7	92.3	0.0	0.0	0.0	12.0	1.0	92.3	7.7
S-17	Healthy tomato 1			0	13	0			0.0	100.0	0.0	0.0	0.0	13.0	0.0	100.0	0.0
S-21	Healthy tomato 2			0	12	1			0.0	92.3	7.7	0.0	0.0	12.0	1.0	92.3	7.7
S-12	ANSV	PV-1027		0	13	0			0.0	100.0	0.0	0.0	0.0	13.0	0.0	100.0	0.0
S-15	CSNV2	PV-0529		0	13	0			0.0	100.0	0.0	0.0	0.0	13.0	0.0	100.0	0.0
S-2	GRSV	PV-0205		0	13	0			0.0	100.0	0.0	0.0	0.0	13.0	0.0	100.0	0.0
S-13	INSV2	PV-0281		0	13	0			0.0	100.0	0.0	0.0	0.0	13.0	0.0	100.0	0.0
S-10	TCSV	PV-0390		0	12	1			0.0	92.3	7.7	0.0	0.0	12.0	1.0	92.3	7.7
S-18	TSWV	PV-1175	1,000,000	1			1	11	7.7	0.0	0.0	7.7	84.6	11.0	2.0	84.6	15.4
S-20	TSWV	PV-1175	1,000,000	1			3	9	7.7	0.0	0.0	23.1	69.2	9.0	4.0	69.2	30.8
S-1	TSWV	PV-1175	100,000	0			1	12	0.0	0.0	0.0	7.7	92.3	12.0	1.0	92.3	7.7
S-19	TSWV	PV-1175	100,000	0			1	12	0.0	0.0	0.0	7.7	92.3	12.0	1.0	92.3	7.7
S-11	TSWV	PV-1175	10,000	0			0	13	0.0	0.0	0.0	0.0	100.0	13.0	0.0	100.0	0.0
S-14	TSWV	PV-1175	10,000	0			0	13	0.0	0.0	0.0	0.0	100.0	13.0	0.0	100.0	0.0
S-3	TSWV	PV-1175	1,000	0			0	13	0.0	0.0	0.0	0.0	100.0	13.0	0.0	100.0	0.0
S-9	TSWV	PV-1175	1,000	0			0	13	0.0	0.0	0.0	0.0	100.0	13.0	0.0	100.0	0.0
S-6	TSWV	PV-1175	100	0			0	13	0.0	0.0	0.0	0.0	100.0	13.0	0.0	100.0	0.0
S-22	TSWV	PV-1175	100	0			2	11	0.0	0.0	0.0	15.4	84.6	11.0	2.0	84.6	15.4
S-8	TSWV	PV-0182	1,000	0			0	13	0.0	0.0	0.0	0.0	100.0	13.0	0.0	100.0	0.0
S-4	TSWV	PV-0182	1,000	0			0	13	0.0	0.0	0.0	0.0	100.0	13.0	0.0	100.0	0.0
S-5	TSWV	PV-0389	100,000	1			3	9	7.7	0.0	0.0	23.1	69.2	9.0	4.0	69.2	30.8
S-16	TSWV	PV-0389	100,000	0			3	10	0.0	0.0	0.0	23.1	76.9	10.0	3.0	76.9	23.1
Total				4	101	2	14	165	1.4	35.3	0.7	4.9	57.7	266.0	20.0	93.0	7.0

INC, inconclusive; TN, true negative; FP, false positive; FN, false negative; TP, true positive.

was not possible to check this for all of the TPS participants separately. Therefore, a fixed Cq cut-off value of 35 was introduced for all of the RT-qPCR datasets except when the Cq cut-off value applied by a TPS participant appeared to be reasonable. This and similar modifications can be carried out by the TPS organiser to respect the 'trueness' of the results, and to keep as much data as possible in the calculations of the statistical parameters, which will provide more reliable measurements of the performances of the tests.

Below, an example from the data analyses of the TPS for TSWV is presented. First, each test was analysed separately (Table 3.10), and then the tests within each method were compared (Table 3.11). The colour coding is consistent across Tables 3.10 and 3.11, with concordant results in green, inconclusive results in yellow, and non-concordant results in red. In this way it is possible to judge which test is better for which purpose. Tables 3.10 and 3.11 can serve as templates for the analysis of the results of other TPS.

To facilitate their comparisons, the results summarising the performances of the different tests can also be presented in graphical forms, using histograms or scatter plots (e.g., see Figs. 3.3, 3.4, and 3.5). Whatever representation is used, it is of utmost importance to be consistent with the data presentation. If the results of tests were obtained based on different sample panels they must not be presented and compared on the same plots, as such representation can be misinterpreted and lead to the disqualification of perfectly good tests. Wrong presentation of the results can lead to the wrong interpretation of the TPS results.

From the data on the diagnostic sensitivity (*DSE*) and diagnostic specificity (*DSP*), it is possible to calculate the likelihood ratios (*LR*). The positive (+) and negative (−) likelihood ratios can be calculated as:

Table 3.11 Example of the comparison of performance parameters determined for individual molecular tests included in the test performance study for tomato spotted wilt orthotospovirus, over all of the submitted datasets

Diagnostic parameter	RT-PCR	RT-qPCR_1	RT-qPCR_2	RT-qPCR_3
Total data sets	13	13	14	13
Expected positives	182	182	196	182
Total data points	286	286	308	286
INC	4	0	3	0
TN	101	100	102	99
FP	2	4	7	5
FN	14	1	1	1
TP	165	181	195	181
INC %	1.4	0.0	1.0	0.0
TN %	35.3	35.0	33.1	34.6
FP %	0.7	1.4	2.3	1.7
FN %	4.9	0.3	0.3	0.3
TP %	57.7	63.3	63.3	63.3
Concordant	266	281	297	280
Non-concordant	20	5	11	6
Concordant %	93.0	98.3	96.4	97.9
Non-concordant %	7.0	1.7	3.6	2.1
Diagnostic sensitivity %	92.2	99.5	99.5	99.5
Diagnostic specificity %	98.1	96.2	93.6	95.2
False positive rate %	1.9	3.8	6.4	4.8
False negative rate %	7.8	0.5	0.5	0.5
Accuracy %	93.0	98.3	96.4	97.9
Power %	90.7	99.5	99.5	99.5
Positive predictive value %	98.8	97.8	96.5	97.3
Negative predictive value %	87.8	99.0	99.0	99.0
Diagnostic odds ratio	595.2	9050.0	5682.9	7167.6

INC, inconclusive; TN, true negative; FP, false positive; FN, false negative; TP, true positive.

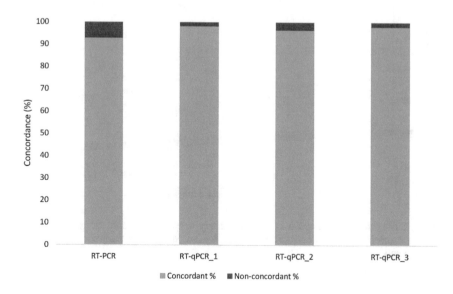

Fig. 3.3 Histogram showing concordance result rates produced by the molecular tests included in the test performance study for tomato spotted wilt orthotospovirus

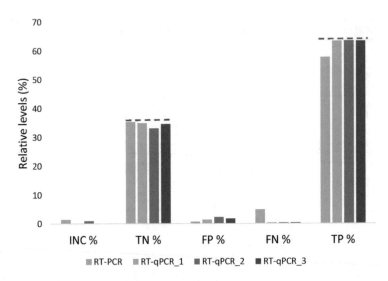

Fig. 3.4 Histogram showing inconclusive samples (INC), true negatives (TN), false positives (FP), false negatives (FN) and true positives (TP) for the molecular tests (as indicated) included in the test performance study for tomato spotted wilt orthotospovirus

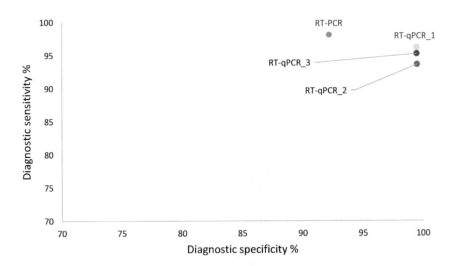

Fig. 3.5 Graphical representation of the relationship between the diagnostic specificities and diagnostic sensitivities for the molecular tests in the test performance study for tomato spotted wilt orthotospovirus

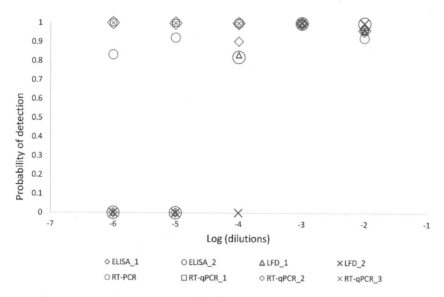

Fig. 3.6 Graphical representation of the analytical sensitivities (as probabilities of detection) for all of the tests included in the test performance study for tomato spotted wilt orthotospovirus. For this analysis, the results of the dilution series of tomato spotted wilt orthotospovirus PV-1175 were used

$$LR+ = DSE / (1-DSP)$$
$$LR- = DSP / (1-DSE)$$

The greater the *LR+* (or *LR–*) for any particular test, the more likely that a positive (or negative) test result is a true positive (or true negative). Likelihood ratios of >10 are considered to be indicative of highly informative (and potentially conclusive) tests.

From the decision-making point of view, using tests with very high negative likelihood ratios means that infected material is less likely to be released (high *LR–* ≡ lower chance of releasing infected material). On the other hand, a high positive likelihood ratio means that for a positive result, the chance that the sample is actually infected is high. Further practical information is available on the YouTube channel in a video entitled: "The analysis of TPS results - video5 - Extending analysis to other performance criteria"; link: https://youtu.be/otDdi5sY_uU.

There are several possible ways for the visualisation of the analytical sensitivity. An example is shown here in Fig. 3.6.

Among the basic requirements for tests used in diagnostics, they need to be repeatable and reproducible. In the context of a TPS, repeatability is the probability of obtaining the same result (positive or negative) from repeated samples analysed in the same laboratory, while reproducibility is the probability of obtaining the same result from repeated samples analysed across different laboratories (EPPO PM 7/122 (1)).

Table 3.12 Summary of the repeatability (%) of tests during the test performance study for tomato spotted wilt orthotospovirus

Test	L01	L02	L03	L04	L05	L06	L07	L09	L10	L11	L12	L13	L14	L15	L17	L18	L19	L20	L21	Average
ELISA_1			100	100		100	100	77.8	100	100	100	100	100	100	88.9	100	100	66.7	100	95.8
ELISA_2			100	100			100	88.9	100	100	100			100	88.9	100	100	88.9	100	97.6
LFD_1			100	100	100		100	100	100		100		77.8		100	100	100	100	100	98.3
LFD_2			88.9	100	100		100	100	100		100				100	100	100	100	100	99.1
RT-PCR	100		■	100			77.8	88.9	100	100	100		■		77.8		100	77.8	100	92.3
RT-qPCR_1	88.9			88.9	100		77.8	■	100	100	100	100		100	100		100	100	100	96.6
RT-qPCR_2	88.9			88.9	100		77.8	88.9	100	100	100	100		100	100		100	88.9	100	94.8
RT-qPCR_3	88.9			88.9	100		77.8	■	100	100	100	100		100	100		100	100	100	96.0
Average	91.7		97.8	93.1	100	100	88.9	90.7	100	100	100	100	88.9	100	94.4	100	100	90.3	100	

Value (%): 100, 90, 80, 70, 60, 50

■ Outlier　　□ Not participated

In the TPS for TSWV, the repeatability was calculated taking onto account the number of samples for which both replicates gave the same result, divided by the total number of samples analysed in the laboratory. For this calculation of the repeatability, samples with other orthotospoviruses were excluded because they were only tested in single replicates. For all tests included in the TPS, the repeatability was high, with a range from 92.3% to 99.1% (Table 3.12). At the laboratory level, the repeatability was from 88.9% to 100% (Table 3.12).

The reproducibility was calculated according to the number of recurrent results per sample, divided by the total number of results per sample. Overall, the reproducibility for the tests included in the TPS were high, with a range from 92.9% to 98.5% (Table 3.13). At the sample level, the lowest reproducibility was for a non-target orthotospovirus isolate when tested with a test that might have weak cross-reactivity (e.g., a GRSV isolate tested with the ELISA_1 was positive according to half of the participants in the TPS, while half of the participants reported it as negative) (Table 3.13). These data demonstrate that it is important that the TPS includes samples with non-targets that might react in tests that are not highly specific (with their evaluation not only in one laboratory).

Analyses of TPS results should also include a comparison with the results of the preliminary studies. If a significant discrepancy occurs in any of the diagnostic parameters, this should be examined and explained, and the TPS organiser should clearly state whether this discrepancy affects the conclusions of the TPS, and whether the results of the TPS are valid, and to what extent. Some of the possible contributions to the differences in results might relate to different preparation protocols for the samples (e.g., use of slightly different buffers, heat deactivation of the extract, storage of the samples, addition of some chemicals). After careful consideration and examination of possible causes, the TPS organiser should provide a clear explanation of these details in the TPS report. If this happens with a commercial test or kit, open and transparent communication with the manufacturer is essential.

Table 3.13 Summary of the reproducibility (%) of tests during the test performance study for tomato spotted wilt orthotospovirus

Type of sample	Sample	DSMZ code	Dilution (-fold)	ELISA_1	ELISA_2	LFD_1	LFD_2	RT-PCR	RT-qPCR_1	RT-qPCR_2	RT-qPCR_3
Healthy tomato	Healthy tomato 1			93.8	100.0	100.0	100.0	92.3	92.3	92.9	92.3
	Healthy tomato 1			100.0	92.9	100.0	91.7	100.0	100.0	100.0	100.0
	Healthy tomato 2			100.0	100.0	100.0	100.0	92.3	84.6	71.4	84.6
Other orthotospoviruses	ANSV	PV-1027		62.5	85.7	92.3	91.7	100.0	100.0	92.9	100.0
	CSNV2	PV-0529		100.0	85.7	100.0	100.0	100.0	100.0	92.9	92.3
	GRSV	PV-0205		50.0	92.9	100.0	83.3	100.0	92.3	92.9	92.3
	INSV2	PV-0281		93.8	100.0	100.0	100.0	100.0	100.0	100.0	100.0
	TCSV	PV-0390		100.0	85.7	100.0	100.0	92.3	100.0	85.7	100.0
TSWV dilution series	TSWV	PV-1175	1,000,000	93.8	100.0	100.0	100.0	84.6	100.0	100.0	100.0
	TSWV	PV-1175	1,000,000	100.0	92.9	100.0	100.0	69.2	100.0	100.0	100.0
	TSWV	PV-1175	100,000	93.8	100.0	100.0	100.0	92.3	100.0	100.0	100.0
	TSWV	PV-1175	100,000	87.5	100.0	92.3	100.0	92.3	100.0	100.0	100.0
	TSWV	PV-1175	10,000	87.5	85.7	76.9	100.0	100.0	100.0	100.0	100.0
	TSWV	PV-1175	10,000	87.5	78.6	76.9	100.0	100.0	100.0	100.0	100.0
	TSWV	PV-1175	1,000	100.0	100.0	100.0	100.0	100.0	100.0	100.0	100.0
	TSWV	PV-1175	1,000	100.0	100.0	92.3	100.0	100.0	100.0	100.0	100.0
	TSWV	PV-1175	100	100.0	100.0	92.3	100.0	100.0	100.0	100.0	100.0
	TSWV	PV-1175	100	93.8	100.0	100.0	100.0	84.6	92.3	92.9	92.3
TSWV medium concentration	TSWV	PV-0182	1,000	nt	nt	nt	nt	100.0	100.0	100.0	100.0
	TSWV	PV-0182	1,000	nt	nt	nt	nt	100.0	100.0	100.0	100.0
	TSWV	PV-0182	100	100.0	100.0	nt	nt	nt	nt	nt	nt
	TSWV	PV-0182	100	100.0	100.0	nt	nt	nt	nt	nt	nt
	TSWV	PV-0182	10	nt	nt	100.0	100.0	nt	nt	nt	nt
	TSWV	PV-0182	10	nt	nt	100.0	100.0	nt	nt	nt	nt
TSWV at concentration close to limit of detection	TSWV	PV-0389	100,000	nt	nt	nt	nt	69.2	100.0	100.0	100.0
	TSWV	PV-0389	100,000	nt	nt	nt	nt	76.9	100.0	100.0	100.0
	TSWV	PV-0389	10,000	100.0	100.0	nt	nt	nt	nt	nt	nt
	TSWV	PV-0389	10,000	100.0	100.0	nt	nt	nt	nt	nt	nt
	TSWV	PV-0389	1,000	nt	nt	100.0	100.0	nt	nt	nt	nt
	TSWV	PV-0389	1,000	nt	nt	100.0	100.0	nt	nt	nt	nt
Average				92.9	95.5	96.5	98.5	93.0	98.3	96.4	97.9

Value (%)
100
90
80
70
60
50

nt, not included in the sample panel

TSWV, tomato spotted wilt orthotospovirus; ANSV, alstroemeria necrotic streak orthotospovirus; CSNV, chrysanthemum stem necrosis orthotospovirus; GRSV, groundnut ringspot orthotospovirus; INSV, impatiens necrotic spot orthotospovirus; TCSV, tomato chlorotic spot orthotospovirus

3.8 Reports and Dissemination Activities

After the organiser of a TPS has carried out the analysis of the data from the TPS, the reports should be distributed to the participants, and results can be published in scientific articles or presented in different conferences, stakeholder meetings, etc. If the TPS included commercial kits/tests, it is recommended that the organisers of a TPS arrange multilateral and/or bilateral meetings with the producers of any commercial tests or kits included, to discuss the results obtained with them prior to finishing the TPS report. The aim of such a meeting is to communicate the results obtained to the producer (company, or in some cases scientific organisation producing the test), to discuss the performances of their tests, and to understand the reasons of unexpected results (if any). If the TPS organiser modified the manufacturer's instructions (e.g., change of matrix, chemicals), given these altered conditions of use of the kit, the TPS organiser and the producers of commercial tests or kits might consider whether the name of the commercial tests or kits should be mentioned in the final report or disseminated by other means. If it is decided that the brand name should not be published, the results for the test can still be published anonymously. Alternatively, the TPS organiser should clearly state that the conditions were different from those recommended by the company, and that the results do not reflect the use as recommended by the company. However, if the commercial tests were used according to the producer recommendations, the results of the tests should be available publicly with the commercial name of the test or kit, because of their importance for the whole diagnostic community. Companies will anyway benefit from receiving detailed information about the performance of their tests (sometimes under different conditions, with different chemicals), while TPS organisers will be able to better understand the results obtained and their possible variations, as well as the impact of each step/modification on the final result. These meetings contribute to better understanding of the mutual needs and possibilities, and allow a more accurate presentation of the results.

One of the important goals of this instruction book is the provision of some universal guidelines for the organisation of a TPS, while at the same time taking into account the potential peculiarities of any particular TPS. Therefore, a harmonisation of TPS reports is proposed, to make their preparation easier, to prevent unintentional omission of important information, and to ensure the high quality of the results presented. In addition, this can help to avoid incorrect interpretation of the results by stakeholders.

Organisers of the TPS developed a unique outline for the TPS reports that was tested and was shown to be useful and valuable in the 12 different TPS. The common template consists of a front page where the basic data on the TPS report should be provided, such as the title of the TPS, the organiser and the contact data, and the version of the TPS report. These are followed by a common outline for all TPS. This outline consists of the following main parts: Context of the application; Methodology of the evaluation; Preliminary studies for evaluation of the method performance; The test performance study; Conclusions; and References. An example with this outline is given in Appendix 8. TPS organisers can adapt the outline based on their needs. For all TPS organisers, it is mandatory to have a disclaimer at the beginning of the TPS report, which states that the results presented in the TPS report only reflect the specific case study and the associated performance results of the commercial reagents at the time when they were included in the study. Rapid changes in the field of biotechnology can lead to improvements/changes in the chemicals related to such diagnostics, and therefore, as reagents can change over time and testing conditions are, to some extent, generally specific, it is important to clearly state that such changes can affect the results.

After the TPS is finished, it is important to disseminate the TPS results such that they reach all potentially interested stakeholders. It is also possible to disseminate the information on the organisational part of the TPS, before the TPS is finalised. In this way, duplication of the work can be avoided, which allows a more efficient use of human, technical, material and financial resources. Such dissemination/exchange of the knowledge will contribute to tackling important problems in the field of study.

In this section we present the steps taken in the VALITEST Project to maximise the impact of the work carried out in the framework of the TPS, and to stimulate, optimise and strengthen the interactions between the stakeholders in Plant Health for better diagnostics. During the Project, the partners established better contacts with the scientific community, stakeholders and policy makers to ensure better dissemination of the TPS results. Furthermore, the links that they established with regional and international standardisation bodies enabled widespread dissemination of the validation data obtained by the Project partners, in particular via the freely available and accessible EPPO database on diagnostic expertise (https://dc.eppo.int/). The EPPO database on diagnostic expertise is a large database that contains the inventory of the diagnostic expertise available in the EPPO region and provides access to the validation data for diagnostic tests for regulated pests. In addition, the results of some of these TPS are planned to be incorporated into revised EPPO diagnostic protocols, and to be published in peer-reviewed scientific journals with open access. The results of these TPS were also presented to wide audiences at several conferences (see details in Petter et al. 2021).

Due to the outbreak of the COVID-19 pandemic, which occurred in the middle of the Project, the VALITEST partners adapted all planned training activities to be available through online platforms. This change improved the possibilities to strengthen the network links between the TPS organiser and interested stakeholders, which allowed an even wider audience to be reached. Also, due to the technological advantages, most of these activities will be available indefinitely and will be "only one click away" (e.g., video recordings from webinars that are available on the Project website, at https://www.valitest.eu/training/activities_and_webinars; and on the EPPO YouTube channel (at https://www.youtube.com/playlist?list=PLoVf4 Pt04Db4aCrCOzZ33QMzDEa1eMtYZ).

These strategies deployed during the VALITEST Project can be used by other TPS organisers to achieve better visibility for their results and to reach larger audiences.

Open Access This chapter is licensed under the terms of the Creative Commons Attribution 4.0 International License (http://creativecommons.org/licenses/by/4.0/), which permits use, sharing, adaptation, distribution and reproduction in any medium or format, as long as you give appropriate credit to the original author(s) and the source, provide a link to the Creative Commons license and indicate if changes were made.

The images or other third party material in this chapter are included in the chapter's Creative Commons license, unless indicated otherwise in a credit line to the material. If material is not included in the chapter's Creative Commons license and your intended use is not permitted by statutory regulation or exceeds the permitted use, you will need to obtain permission directly from the copyright holder.

Chapter 4
Conclusions

Ana Vučurović ⓘ**, Géraldine Anthoine, Charlotte Trontin, Tanja Dreo,
Tadeja Lukežič, Françoise Petter, Maja Ravnikar** ⓘ**, and Nataša Mehle** ⓘ

4.1 Main Challenges and Recommendations

The organisation of a TPS is a complex and demanding process. This can be made
easier if timelines, rules and specific criteria are defined early, and are followed. This
requires more preparation work; however, this pays off when the TPS is running. It is
worth trying to foresee possible scenarios and difficulties. In this way, reaction times
to act will be shorter, and will affect the running of the TPS to a lesser extent.
Possible delays need to be taken into account. For example, delays of sample
dispatching due to delays in acquiring Letters of Authorisation can lead to less
time available for the TPS participants to perform the tests. Therefore, the TPS
participants need to be informed in advance if they will need to acquire a Letter of
Authorisation (or other import permits). They also need to be given enough time to

A. Vučurović (✉) · T. Dreo · T. Lukežič · M. Ravnikar
Department of Biotechnology and Systems Biology, National Institute of Biology,
Ljubljana, Slovenia
e-mail: ana.vucurovic@nib.si; tanja.dreo@nib.si; tadeja.lukezic@nib.si; maja.ravnikar@nib.si

G. Anthoine
French Agency for Food, Environmental and Occupational Health and Safety (ANSES), Plant
Health Laboratory, Angers, France
e-mail: geraldine.anthoine@anses.fr

C. Trontin · F. Petter
European and Mediterranean Plant Protection Organization, Paris, France
e-mail: trontin@eppo.int; petter@eppo.int

N. Mehle
Department of Biotechnology and Systems Biology, National Institute of Biology,
Ljubljana, Slovenia

University of Nova Gorica, School for Viticulture and Enology, Vipava, Slovenia
e-mail: natasa.mehle@nib.si

© The Author(s) 2022
A. Vučurović et al. (eds.), *Critical Points for the Organisation of Test Performance
Studies in Microbiology*, Plant Pathology in the 21st Century 12,
https://doi.org/10.1007/978-3-030-99811-0_4

prepare for the TPS, to order the specific chemicals, and to perform the tests, while still leaving some time to repeat some tests if needed.

Even though communication with TPS participants is sometimes time consuming, it is crucial to avoid misunderstandings and to avoid the exclusion of results from the analysis. Open and transparent communication between the organisers of a TPS and the commercial kit providers can be crucial, for similar reasons. This communication can also help with the acquisition of chemicals needed for a TPS by the TPS participants, and in the case of some specific chemicals with short expiry dates (e.g., serological tests), this can prevent a possible shortage.

Based on the results of the analysis, the TPS organiser should draw the appropriate conclusions on the performances of the tests included in the TPS, and should determine if they are fit for purpose, considering the scope of the testing. Also, the TPS organiser should comment on the performances of certain tests, in terms of which uses and conditions they are applicable to, and which they are not.

As mentioned several times, open communication between the TPS organiser and the TPS participants is very important. After the completion of the entire TPS, organisers can circulate a satisfaction survey to solicit feedback from the participants. The participants can be asked to rate the experience of their participation, their satisfaction with the organising process, and the possible areas for improvement. Certainly, such surveys should be confidential, and in line with the General Data Protection Regulations (GDPR for the EU region).

Even though a TPS is not designed as an interlaboratory comparison where the proficiency of the laboratories is evaluated, after each laboratory receives the TPS report, they can analyse their own performances. These data can thus be used to identify whether some correction measurements need to be implemented to improve their performances.

The organisation of TPS also has common elements with the organisation of proficiency testing, such as the preparation of the plan, the communication with the participating laboratories, the preparation of samples, testing for stability and homogeneity, etc. The guidance given in this book can also be useful in this context.

As explained in the Introduction, the offers for TPS are still relatively limited in plant health. The authors hope that by sharing their experience and providing guidance and tips in this book, this will help to increase the numbers of laboratories that are willing to organise test performance studies for the global benefit to the plant health diagnostic community.

4.2 Summary

In this book we have used the collected experience that our TPS organisers gained through the organisation of 12 TPS, as well as previous extensive experience each participating institution gained through organisation of similar studies. This information can be used as a 'textbook' for organisers of future TPS, not only in the field of plant pest detection, but also in other areas of microbiology. The possibility to use the knowledge gathered in this book beyond the field of plant health, will enable the creation of new network connections and exchange, and as an outcome will continuously improve the concept and the organisation of TPS.

The organisation of a TPS and similar studies is a difficult process, with the need for financing and time. Moreover, it requires a high level of expertise from the TPS organiser to ensure a smooth process and reliable results. However, TPS studies represent the best way to obtain the most reliable validation data for a test of interest, and to study the performance of the test in different laboratories under different conditions. This can help to minimise the uncertainties in the performance characteristics of a test. In addition, TPS are also the most transparent approach to discuss tests and testing strategy with different stakeholders. It provides objective evidence for each test evaluated, despite it being carried out under specific conditions.

By organising 12 TPS on 11 prioritised pests, we aimed to reach the widest possible audience of interested stakeholders in the field of plant pest diagnostics. In all, 640 diagnostic laboratories were invited to participate in these 12 TPS, and 242 of these laboratories confirmed their participation. Five companies also participated, where their main activities included the production of commercial tests for plant pests. Furthermore, their engagement in the test selection process enabled better understanding of the possibilities to change some aspects of the test implementation, and prevented significant deviation of the results due to the changes applied. Each TPS organiser also contacted companies that did not directly participate in the Project (i.e., other manufacturers of commercial tests for plant pests), and indeed, some of their tests were also evaluated in the preliminary studies, if not in the TPS. The evaluation process involved open communication with the companies. As the intergovernmental organisation that is responsible for international cooperation in plant protection, EPPO was included to help to establish and ensure certain levels of harmonisation of all of the procedures during the preparation and organisation of these TPS. During the Project, we harmonised the organisation of the TPS as much as possible between the Project partners, with the aim to set the standards for further similar studies. This was enabled and was achieved through the development and use of common criteria and rules for the selection of tests and laboratories, and common templates (e.g., TPS invitation letters, contracts, technical sheets) and common procedures. When we established the possibility during this process to improve the procedures to better serve the purpose (based on the experience gained in the Project), we up-dated some of these documents. As a Project outcome, this book presents the rules, criteria and documents that we developed and that can serve as a blueprint for similar future studies.

Appendices

In the Appendices with the examples of TPS documents, the information that needs to be provided by the TPS organiser is indicted in the form of *[grey letters in italics and between square brackets]*.

© The Editor(s) (if applicable) and The Author(s) 2022

A. Vučurović et al. (eds.), *Critical Points for the Organisation of Test Performance Studies in Microbiology*, Plant Pathology in the 21st Century 12, https://doi.org/10.1007/978-3-030-99811-0

Appendix 1. TPS Invitation Letter

[add logo]

To whom it may concern

[add the date of the sending]

TPS organizer:
[add the name of the TPS organiser]

Pest name:
[add pest name]

Subject: [add the name of the project/study]

Encl.: TPS Interest - Information Form for [add pest name]

Followed by:
[add the name of the person in charge for the TPS organisation]

Dear Sir or Madam,

We are contacting you [briefly explain the framework of the organization and the aim of the study]. Validation is essential to provide information on the performance of the tests that are used in diagnostic. However, most detection and identification tests are currently only validated on an intra-laboratory basis or through limited test performance studies (TPS), and there is a need to further harmonize practices.

Telephone number:
[add telephone number of the coordinator]

E-mail:
[add e-mail address/es of the coordinator]

By this message, we are seeking your interest to take part to a test performance study. Indeed, you have been identified as a potential participant for a TPS for the detection of [add pest name].

The expected timeline is as follows:
Period of TPS: [define the period]
Sending of the samples: [add the time span when samples will be dispatched]

Deadline for performing analysis: [add deadline]
Deadline for participants reporting of on the results: [add deadline]

In the table below you can find the methods to be evaluated together with the scope of TPS [add the scope of the TPS and the name of the pest]:

Method	[add Method 1, e.g. ELISA]	[add Method 2, e.g. PCR based methods]	.....	[add Method n]

			.....	
Sample type	*[describe the sample type, e.g., infected/ non-infected plant material]*	*[describe the sample type, e.g., infected/ non-infected plant material]*		*[describe the sample type, e.g., infected/ non-infected plant material]*
Matrix	*[define the matrix, e.g., leaves, seeds, plant species, etc.]*	*[define the matrix, e.g., leaves, seeds, plant species, etc.]*	.....	*[define the matrix, e.g., leaves, seeds, plant species, etc.]*
Suitable for	*[define the intended use, e.g., symptomatic/ asymptomatic samples]*	*[define the intended use, e.g., symptomatic/ asymptomatic samples]*	.....	*[define the intended use, e.g., symptomatic/ asymptomatic samples]*
Purpose	*[define the intended purpose of use, e.g., detection/ identification/ detection and identification, etc.]*	*[define the intended purpose of use, e.g., detection/ identification/ detection and identification, etc.]*	.....	*[define the intended purpose of use, e.g., detection/ identification/ detection and identification, etc.]*
Type of controls needed	*[describe the controls which will be needed, e.g., commercial positive control (PC), negative control (NC), buffer control (BC)]*	*[describe the controls which will be needed, e.g., negative amplification control (NAC), internal positive control (e.g., primers amplifying nad5)]*	.....	*[describe the controls which will be needed]*
No. of samples	*[add number of samples]*	*[add number of samples]*	.....	*[add number of samples]*
Maximal number of tests to be evaluated[1]	*[add number of tests]*	*[add number of tests]*	.....	*[add number of tests]*

[1]Please note that the number of methods and tests is indicative at this step. It will be adjusted according to the results of preliminary tests and participation interest. Maximum number of tests in TPS will be *[add number of tests]*.

[define the financial requirements for the participation. Example: If the participating laboratory is not a member of the [add the name of project consortia/ organisation], it is informed that participation in the TPS will be at its own cost (i.e., consumables, chemicals, reagents) as the project budget does not provide funding to organisations outside the project consortium. If the participating laboratory is a member of the [add the name of the project consortium/ organisation] consortium, it is informed that it already benefits from a dedicated budget for participation in the TPS.]

As a participant to the TPS, you would receive the evaluation report and you would be associated to the results exploitation. The samples are expected to include *[describe the type of material, e.g., leaf material, plant extract, etc.]*.

Participants will be selected based on pre-defined criteria. In order to optimize our organization and the reliability of the TPS, we would like to get some practical details concerning your organization.

To express your interest in participating, please, fill in the enclosed excel file named 'TPS Interest - Information Form' and return it **in digital form** by e-mail to *[add email address 1 of TPS organiser]* **and** *[add email address 2 of TPS organiser]* by *[add deadline]*. We will inform you about your possibility to participate by *[add deadline]*. There are critical selection criteria, which need to be fulfilled for participation in TPS:

- Time schedule described above compatible with your availability
- Authorization by the national competent authority to work with *[add pest name/s]*, **etc.** (*[define the type of material which will be shipped to the participants]*)
- Traceability in place / QA in place
- Possibility to obtain an import document or Letter of Authority (EU countries)

Candidates which are able and committed to perform all methods will have an advantage in TPS selection process, whereas it is necessary to perform all tests for the selected method. If more than the maximum number of laboratories fulfill all selection criteria for a specific test, participants are selected on "first come, first served" basis.

Please do not hesitate to contact us should you require additional information.

Yours sincerely,

[add full name and signature of the person in charge for the coordination]

Appendix 2. TPS Participant's Contract

[add logo]	**Test Performance Study (TPS)** **Participant's Contract**

To register for the Test Performance Study (TPS),
please fill in this form and sign and return it by email
(together with the *[specify, e.g., Excel]* **file with your contact information) to:**
[add e-mail address 1 of TPS organiser]
[add e-mail address 2 of TPS organiser]

Identification of the TPS

TPS code: *[add unique code of the TPS]*

Laboratory (short name)	
Laboratory (full name)	

Registration

The laboratory agrees to participate in the TPS: ☐ Yes [1]

[1] By validating its registration, the laboratory identified in this contract (henceforth referred to as the "participating laboratory") agrees to participate in the TPS organised by *[add full name of the TPS organiser]* under the conditions of participation described below.

Please provide all of the information requested in the table below:

	Select the method that will be tested in your facilities:			
Analytical tests to be evaluated	**Methods for** *[add pest name]* **Indicate the methods to be evaluated by ticking the relevant box** For the benefit of the project and to get enough data, the participant is invited to **evaluate all of the tests related to the methods it is willing to perform.**			
	☐ *[add method 1]*	☐ *[add method 2]*	☐ ...	☐ *[add method n]*
	• *[add test 1]* • *[add test 2]* ... • *[add test n]*	• *[add test 1]* • *[add test 2]* ... • *[add test n]*	...	• *[add test 1]* • *[add test 2]* ... • *[add test n]*
	The participating laboratory will have to apply the detailed protocol of each test provided as Appendices to the TPS Technical Sheet. *[briefly explain how the protocols are prepared;*			

	e.g., "These protocols are adapted from the original publications", or refer to the manufacturer's instructions.] Any comments: _____
Requirements	**The regulatory status of the samples with pest that are to be dispatched:** Type of material: *[explain the type of material, e.g., "plant extract prepared from frozen leaves of tomato potentially infected with tomato spotted wilt orthotospovirus and other tospoviruses"].* Quantity of material: *[define the maximal number of sample panels a participant can receive]* panels of samples and controls (for participants implementing all of the methods), each containing: *[explain in detail how the samples will be packed (e.g., tubes, extraction bags, vials, etc.); provide the exact number of samples and quantities of the materials that the participant will receive (e.g., 24 safe-lock 2 mL tubes with 100 µL plant leaf extract in each)]* Scientific name of the material, including the pest(s) concerned: *[provide the scientific names for the materials that will be sent to the participants, including the names of the pest(s) concerned; any closely related pests that might be included in the samples; the scientific names of the plants which were used for the preparation of the samples and that might be part of the sample panels]* Sender of the panel: *[add full name of the TPS organiser]* *[add Department or organisational unit of the TPS organiser-if applicable]* *[add the full postal address of the TPS organiser]* *[add the number of sample panels]* separate sample panels (including controls) that will be sent to the participants:

Sample/ control designation	Type	*[add e.g., Volume of the plant extract/ RNA in the samples (in µL)]*			
		[add method 1]- [add test 1]	*[add method 2]- [add test 2]*	......	*[add method x]- [add test y]*
S-1	Test item *[describe*]*	*[add volume/amount]*	*[add volume/amount]*	....	*[add volume/amount]*
S-2	Test item *[describe*]*	*[add volume/amount]*	*[add volume/amount]*	....	*[add volume/amount]*
S-3	Test item *[describe*]*	*[add volume/amount]*	*[add volume/amount]*	....	*[add volume/amount]*
....	....	....	....	....	....
S-*n*	Test item *[describe*]*	*[add volume/amount]*	*[add volume/amount]*	....	*[add volume/amount]*
C	Control *[describe*]*	*[add volume/amount]*	*[add volume/amount]*	....	*[add volume/amount]*
....	....	....	....	....	....

[*define the content, e.g., plant extract, plant material, RNA, DNA, etc.]

Regulatory requirements concerning the plant health regulations of your country (territory) to allow the sending of the samples [tick the appropriate box]:

☐ No requirements

☐ Requirements[2] : ☐ the parcel must be accompanied by a *[add the name of required import permit, e.g., LoA[3]]*

☐ other, please detail : ▓▓▓▓▓▓▓▓▓▓▓

[2] **IMPORTANT**, if requirements are specified and if the TPS organiser does not have the necessary documents before *[provide the deadline]*, the legislation will not allow the parcel of samples to be sent, and the participation of the laboratory will be compromised.

[3] *[add the name of required import permit, e.g., Letter of Authority]* authorising the circulation of a regulated pest in the *[add the name of the country, union or geographic region, to which the permit is applicable, e.g., European Union]*.

Samples reception requirements

The samples will be dispatched on *[provide the date/ timeframe when the samples will be dispatched]*. The participating laboratory will be available to receive the samples from *[provide the timeframe when the participating laboratories will receive the samples]*.

☐ Yes

☐ No[4]

[4] If no, the participating laboratory must inform the TPS organiser by email: *[add e-mail address 1 of the TPS organiser]*, cc to *[add e-mail address 2 of the TPS organiser]*

Conditions of participation

1) Implementation of the TPS:

The participating laboratory agrees to perform analyses in its laboratory according to the instructions of the TPS organiser (the TPS Instruction Sheet will be sent together with the panel of samples), and under its usual conditions of work.

Any modifications should be reported to the TPS organiser. This kind of information can be very important for the interpretation of the results.

The participating laboratory agrees to communicate any difficulties encountered during the implementation of the TPS.

The participating laboratory agrees to provide all of the results on the TPS Results Form provided to it within the deadline indicated on the TPS Technical Sheet.

[define the financial requirements for the participation. Example: If the participating laboratory is not a member of the [add the name of project consortia/ organisation], it is informed that participation in the TPS will be at its own cost (i.e., consumables, chemicals, reagents) as the project budget does not provide funding to organisations outside the project consortium.

If the participating laboratory is a member of the [add the name of the project consortium/ organisation] consortium, it is informed that it already benefits from a dedicated budget for participation in the TPS.]

2) Transmission of the results:

The participating laboratories are informed that the TPS results will be analysed anonymously. The TPS Report will be transmitted to each participating laboratory in electronic format on *[add the date when the TPS Report will be transferred to the participants].*

For the distribution of the proficiency testing report to the participating laboratories, the following convention of proof applies.

The TPS Report will be transmitted in a protected pdf format attached to an e-mail sent to the two electronic addresses communicated in writing by the participating laboratory in the participant's contract. The signatures of the authorised persons on the first page of the proficiency testing report constitute proof of its validation and its authenticity.

The TPS organiser keeps the original file (electronic version and paper version) and also the evidence of its authenticity.

Moreover, the participating laboratory:

-agrees to not modify this file;

-is informed that the paper editions from the transmitted pdf file are under its sole responsibility;

-recognizes the validity and the convincing strength of this file.

3) **Confidentiality and cooperation:**

The TPS organiser will provide each participating laboratory with *[define the number of sample panels the participant can receive]* panels of coded samples. The coding of the samples will be kept confidential by the TPS organiser until the end of the TPS.

Each participating laboratory agrees not to communicate with the other participants or any third parties regarding the samples or any part of the results.

The participating laboratory is informed that the TPS results might be used anonymously for scientific purposes. The participating laboratory will be associated with exploitation of the results. The participating laboratory agrees not to publish its results in any kind of format or by any means.

The TPS organiser is committed to keeping your information secure and strictly confidential, and not sharing it with any third parties.

The participating laboratory agrees to the handling of its data for the purposes of the TPS in accordance with the General Data Protection Regulation (GDPR). The way in which its data is used for the purposes of the TPS is described in the Appendix.

4) **Confinement:**

The participating laboratory must have adopted measures to prevent the risk of unintentional release of plant pests for which it is approved (e.g., containment, processing of waste, etc.).

The participating laboratory agrees to inform the TPS organiser in the TPS participant's contract of the plant health regulatory requirements of its country (territory), to allow the sending of the samples in conformity with its regulations. The participating laboratory also agrees to complete the necessary formalities to allow the receipt of the samples, and so to ensure its participation in the TPS.

Samples sent by the TPS organiser must be used only in the framework of *[specify, e.g., this TPS study]*.

The TPS organiser declines any responsibility for the use of the samples or any remainders outside the framework of the *[add the name of the project]* project activities.

All samples should be destroyed after submission of the results.

Name:	Date:	Signature:
Function:		

Appendix: GDPR Disclaimer

1 Categories of personal data
Your consent to the processing of personal data is given freely. By registering, you agree that *[add the full name (short name) of the TPS organiser]* will process the following personal data for the purposes of this consent form: title, name, surname, address, e-mail address and your consent to the processing.

2 Safety of important personal data
[add the short name of the TPS organiser] will implement the appropriate technical and organisational measures to protect your important personal information.

3 Retention of the personal data
[add the short name of the TPS organiser] can store your personal data for as long as necessary for the purposes for which the personal data are processed, or for such a time and for such purposes as required or permitted by the applicable law(s). After this period, *[add the short name of the TPS organiser]* will immediately delete or render your personal data anonymous.

4 Your rights
In accordance with applicable law(s) (including exceptions or derogations from this legislation), you have the right to request access to your personal data as well as its rectification, removal or restriction of processing, the right to object to the processing, the right to transfer the data, the right to withdraw consent for the processing of personal data for a specific purpose, if such consent has been previously given, the right to bring a complaint to the supervisory authority in connection with the processing of personal data. Any request has to be notified by email (*[add e-mail address 1 of the TPS organiser]*, *[add e-mail address 2 of the TPS organiser]*).

5 Applicable law(s)
This form is subject to the legislation of the *[specify, e.g., EU]* and is interpreted in accordance with it. This does not affect your legal rights.

Appendix 3. TPS Participant's Contact Information

[add logo]

Test Performance Study (TPS) participant´s contact information
(this document has to accompany participant´s contract)
Please fill in grey boxes and send back [specify, e.g.: *the Excel file in a digital form*] *to:*
[add e-mail address 1 of the TPS organiser]
[add e-mail address 2 of the TPS organiser]

Identification of the TPS

TPS code: [add unique code of the TPS]

Laboratory (short name)

Shipping address
Laboratory (full name)
Street address (Line 1)
Street address (Line 2)
ZIP Code
City
State/Province/Region
Country

Contact
Title (Mrs./Mr.)
First name
Family name
E-mail address
Telephone
Function

Alternative contact
Title (Mrs./Mr.)
First name
Family name
E-mail address
Telephone
Function

Appendix 4. TPS Technical Sheet

[add logo]	**Test Performance Study (TPS) Technical Sheet**

Identification of the TPS

TPS code: *[add unique code of the TPS]*

	Methods included in the TPS for the detection of *[add pest name]*			
	[add Method 1, e.g. ELISA]	*[add Method 2, e.g. PCR based methods]*	.....	*[add Method n]*
Sample type	*[describe the sample type, e.g., infected/ non-infected plant material]*	*[describe the sample type, e.g., infected/ non-infected plant material]*	.....	*[describe the sample type, e.g., infected/ non-infected plant material]*
No. of samples	*[add number of samples]*	*[add number of samples]*	.....	*[add number of samples]*
Number of tests to be evaluated	*[add number of tests]*	*[add number of tests]*	.....	*[add number of tests]*
Number of panels of samples provided per method	*[add number of tests]*	*[add number of tests]*	.....	*[add number of tests]*
Provision of controls by the organiser	*[describe the controls which will be provided by the organiser, e.g., positive control (PC), negative control (NC)]*	*[describe the controls which will be provided by the organiser, e.g., negative isolation control (NIC), positive isolation control (PIC), positive amplification control (PAC)]*	.....	*[describe the controls which will be provided by the organiser]*
Other controls needed (not provided by the organiser)	*[describe the controls which should be provided by the participants, e.g., commercial positive control (PC), negative control (NC), buffer control (BC)]*	*[describe the controls which should be provided by the participants, e.g., negative amplification control (NAC), internal positive control (e.g., primers amplifying nad5)]*	.....	*[describe the controls which should be provided by the participants]*

1. Introduction

The test performance study (TPS) is performed in the framework of *[add project name]*. *[briefly explain project aim and/or scope, e.g., production of new validation data through a test performance study]*.

[add full name of the institution, Department (if applicable)] is organising a TPS for *[specify, e.g., the detection and identification of [add pest name]]*. *[add other details. Example: Each participating laboratory to the TPS has been selected to evaluate the methods which they expressed willingness to perform in the interest form.]* The TPS Technical Sheet will explain the practical details of the organisation and the participation in the TPS.

2. Organisation

The organisation of the TPS is ensured by *[add full name of the institution, Department (if applicable)]*

and is coordinated by:

[add full name of the person in charge for the coordination]
Email: *[add e-mail address/es of the coordinator]*
Telephone: *[add telephone number of the coordinator]*

with the technical expertise of *[if needed add names of the involved personnel]*

3. Participation

3.1. Coding of TPS participants and confidentiality

The TPS organiser will provide each participating laboratory with a panel of coded samples. Coding of the samples will be kept confidential by the TPS organiser until the end of the TPS. The participating laboratories are informed that the TPS results will be analysed anonymously and then each participating laboratory will be given a copy of its own results.

3.2. Resources required for the TPS

3.2.1. Personnel (days):

[provide an estimation of the time needed by the TPS participant to perform the analyses. Here we provide an example of a TPS with eight tests:
The working time for planning and performing the analyses, and submitting the results is estimated at 8 days for one person if the laboratory decides to implement all of the methods included in the TPS.
The estimated times required for performing the analyses are:
-      *1 day for planning and submitting the results*

-      *3 days for ELISA (two tests)*
-      *0.5 day for RNA extraction using column-based kit (24 samples)*
-      *1 day for conventional RT-PCR (one test)*
-      *2 days for real-time RT-PCR (three tests)*
-      *0.5 day for lateral flow device tests (two tests)].*

3.2.2. Consumables:

[provide the information on the consumables needed to perform the analyses. Provide the information if the TPS organiser has negotiated discounts with commercial kit providers for potential TPS participants, and how the discount can be obtained. Provide the information on which chemicals the TPS organiser used for the preliminary study. Provide the information on the amounts of chemicals the participant will need to perform certain tests (this can be presented in the separate table-see example below]

List of critical reagents needed by the participants to implement the TPS.

Method	Number of tests to be evaluated	Number of reactions per test	Reagents
[add Method 1, refer to appendix with details]	*[provide No. of tests]*	*[provide No. of samples, repetitions and controls]*	*[provide names, producers, and lot numbers for all chemicals and/or kits needed to perform the analyses]*
[add Method 2, refer to appendix with details]	*[provide No. of tests]*	*[provide No. of samples, repetitions and controls]*	*[provide names, producers, and lot numbers for all chemicals and/or kits needed to perform the analyses]*
...			

Note: It could be relevant to order more reagents than strictly necessary to have the possibility to perform the analyses again should there be any problems.

[add disclaimer on the use of chemicals if needed, e.g., **The use of this kit implies no approval of them to the exclusion of others that may also be suitable.** *Include justification for why some chemicals were selected and for deviations from manufacturer's protocols, if applicable.]*

[provide the information on which chemicals will be provided by the TPS organiser and under what conditions, e.g., primer-probe mixes, primer mixes (provision of primers and probes for participants usually will not cause significant additional cost for the organiser, but can lower the cost of participation for the laboratories and provide more potential participants)]

3.2.3. Equipment and materials

Standard laboratory equipment is needed.
[provide the information on which main laboratory equipment was used by the TPS organiser in the preliminary study, e.g., which PCR and real-time PCR systems, absorbance readers, etc.]

The use of this equipment implies no approval of them to the exclusion of other equipment that might also be suitable. Experimental protocols presented in the Appendices can be adjusted to the standards of the individual laboratories (e.g., use of other equipment than that indicated), provided that the equipment is adequately validated.

3.3. Special precautions to be taken by the TPS participants

[add text if relevant. Example: When carrying out the tests, participating laboratories will have to make the necessary arrangements to prevent the risk of unintentional release of plant pathogens they might be handling.
After performing the test:
-any remains of the samples and controls are to be destroyed by autoclaving or by any other efficient means for inactivating viruses;
-consumables in contact with the samples or controls are to be autoclaved (tubes, microtitre plates, PCR plates, etc.);
-materials in contact with samples and controls are to be disinfected.]

4. Planning of the TPS

[provide the timeline of the TPS for participants. An example is given in the table below]
Planning for the TPS round:

Steps	Deadlines
Call for participants - contract sending	*[add deadline]*
End of registration	*[add deadline]*
Deadline for regulatory requirements submission	*[add deadline]*
Sample dispatch	*[add deadline]*
Deadline for performing analyses	*[add deadline]*
Deadline for submitting results	*[add deadline]*
Final report transmitted to the TPS participants	*[add deadline]*

5. Communication with participants

5.1. Information provided to the participants

[provide detailed information for the participants on the actions that they need to carry out, and on the actions that will be carried out by TPS organiser. An example is given in the table below]
Detailed information provided to the TPS participants on the actions that they need to carry out for the different steps of the TPS.

TPS step	Documents provided to the participants	Actions to be carried out by the participants
Call for participants - contract sending	-the TPS Technical Sheet -the participant's contract	-Registration, by sending back the TPS participant's contract. -Completion of the regulatory formalities to allow the samples to be sent by the TPS organiser in accordance with the phytosanitary regulations (e.g., request for a

		Letter of Authorisation, and its signing and submission to the TPS organiser)
Sample dispatch	-the TPS Instruction Sheet -acknowledgement of receipt form -the TPS Results Form -the regulatory documents provided by the participant allowing the samples to be sent by the TPS organiser in accordance with the phytosanitary regulations	On receipt of the package: -check the contents of the package -send back the acknowledgement of receipt form Manage the samples, perform the analyses and record the results according to the TPS Instruction Sheet Send back the TPS Results Form, once completed
Sending of the TPS Report	-the TPS Report	

5.2. Communication

For exchanges by e-mail, the participants are asked to always send their e-mails to the two following electronic addresses:

[add e-mail address 1 of the TPS organiser]

and

[add e-mail address 2 of the TPS organiser]

6. Tests to be evaluated

The Appendices give the protocols for the tests that will have to be implemented by the participating laboratory for each method for which it is registered in the TPS. *[add other text, if relevant. Example: These protocols are adapted from the original publications or refer to the manufacturer's instructions.]*

[add, if needed, any additional information for the participants, e.g., IMPORTANT: For the benefit of the project and to obtain enough data, each participating laboratory is invited to evaluate all of the tests related to the methods it is willing to perform for the TPS.]

[tests to be evaluated in the TPS can be presented as in the example table below.]

Tests to be evaluated in the TPS *[specify]*

	Tests for *[pest name]*			
	[add Method 1]	*[add Method 2]*	...	*[add Method n]*
Tests to be evaluated	• *[add test 1 and the number of the Appendix that contains the details of how the test should be performed]* • ...	• *[add test 1 and the number of the Appendix that contains the details of how the test should be performed]* • ...	...	• *[add test 1 and the number of the Appendix that contains the details of how the test should be performed]* • ...

7. Samples to be analysed and sample dispatch to participants

7.1.   Composition of the panel of samples to be analysed

[add the information on what the sample panel contains, including the number of samples (their designation), the number of controls (their designation), how the samples are packed, the tubes, bags, vials, etc.]

[explain in detail the conditions in which the samples should be stored upon their receipt, e.g., -20 °C (frozen), -80 °C, +4 °C, room temperature, etc.]

[explain in detail how many samples will be provided to the participant, and how they will be designated for each method. Example: Several panels of samples will be provided to each participant. For example, a participant registered to implement all of the methods will receive two panels of samples to implement the ELISA tests, two panels of samples dedicated to the on-site detection methods, and a single panel of samples to implement all of the molecular tests. Each panel of samples will be clearly marked.]

[The composition of the sample panels can be provided in the table.]

7.2.   Validation of the samples

The TPS organiser will ensure that the samples used for the evaluation by the laboratories are sufficiently homogeneous and stable.

7.3.   Coding of the samples

The samples will be coded to ensure the full blind testing of the samples. The individual code of sample will be randomly assigned and will be different for each panel of samples provided, and for each participant.

7.4.   Transport

The samples, controls, *[and other TPS material provided by the TPS organiser, if applicable]* will be transported to the participating laboratories *[specify, e.g., on dry ice by a courier (e.g., DHL)]*. If any problems occur concerning the samples, controls *[and/or other TPS materials provided by the TPS organiser, if applicable]*, the participating laboratory will inform the TPS organiser within *[specify, e.g., 24 hours]* of receiving the TPS materials.

7.5.   Sample storage, analysis and result recording

The TPS Instruction Sheet will be sent to each participant, together with the package. It will provide and explain the instructions to be followed:
- once the package is received;
- on the storage of the samples, controls *[and/or other TPS material provided by the TPS organiser, if applicable]*;
- for performing the analyses;
- for recording and submitting the results.

8. Submission of the results

The participants are asked to respect the deadlines for performing the analyses and for submission of the results (provided in the section on Planning of the TPS). The participants are also asked to use the TPS Results Form provided to them for the submission of their results. Once completed, the TPS Results Form is to be returned by e-mail to the two electronic addresses mentioned in the communication section.

9. Processing of the results

The performance of the tests will be evaluated according to the qualitative results submitted by the participating laboratories. Details of the statistical processing of the results will be provided in the TPS Report.

10. TPS Report

The TPS Report will present the results of the participating laboratories anonymously. Each participating laboratory will be sent a copy of its own results. The TPS Report will be transmitted to each participating laboratory by electronic format (protected pdf format) to the two electronic addresses communicated in the participant's contract.

11. References

[provide the list of references]

Name and Signature of the TPS organiser
[add the name and signature of the person responsible for the organisation of the TPS]

[add date]

LIST OF APPENDICES
[provide the list of Appendices. For example:
APPENDIX 1 – test 1
APPENDIX 2 - test 2
APPENDIX ... - ...
APPENDIX n – test n]
[For all of the tests, the Appendices should contain all of the information needed to perform the analyses, including the number of replicates that should be performed, and if applicable, the reprocessing of the samples and controls for the tests upon receiving them from the TPS organiser. Implementation of the tests and interpretation of results, including verification of the controls, should be clearly defined, e.g., as for tests in the Appendices of the EPPO PM7 standards]

Appendix 5. TPS Instruction Sheet

TPS Instruction Sheet

[add logo]

Identification of the TPS

TPS code: *[add unique code of the TPS]*

---

1) Receipt of the parcel

Check the content and conformity of the package, and complete the acknowledgement of receipt.

The package contains up to *[add the number of sample panels]* panels of samples. Each participant will receive the number of panels specified in its registered participation. Thus, a participant registered to implement all of the methods will receive:

-*[add the number of sample panels]* panels of samples to implement *[add method 1 and the number of test within method 1]* tests,

-.....

-*[add the number of sample panels]* panels of samples to implement *[add method n and the number of test within method n]* tests,

Each panel includes *[add the number of samples]* samples and *[add the number of controls]* controls packaged in *[explain in detail how the samples and controls are packaged, e.g., tubes, extraction bags, vials, etc.]*. Panels of samples for the participating laboratories are coded according to the following principle:

-**Panel code (ID of the participant):** in the form "LXX" where XX is a two-digit number

In each panel, the samples are coded according to the following principle:

-**Sample code:** S-X to XX where X is a number from 1 to *[add total number of samples in the panel]*.

In addition, the label on each sample contains the data for which the particular test sample is designated (i.e., the test to be performed), the name of the project, and TPS code (*[add unique code of the TPS]*).

Test to be performed	Designation on the label
[add method 1]	*[add designation on the label for test 1]*
	[add designation on the label for test 2]
	[add designation on the label for test n]
[add method n]	*[add designation on the label for test 1]*

[add designation on the label for test 2]
[add designation on the label for test n]

-Control code: *[define how controls are marked; e.g., positive control, PC; negative control, NC; positive isolation control, PIC; negative isolation control, NIC; positive amplification control, PAC]*

Examples of labels for samples and controls in a panel:

"Side" label for samples   "Top" label for samples

[add test name]

S-X

[add test name]

X is a number from 1 to *[add total number of samples in the panel]*

"Side" label for control   "Top" label for control

[add test name]

PC

[add test name]

[If relevant, also include details of the reagents/chemicals provided, including how they are labelled]

The provided acknowledgement of receipt form (*[add the name of the document]*) should be completed and sent by email to *[add e-mail address 1 of the TPS organiser]* and *[add e-mail address 2 of the TPS organiser]*.
IMPORTANT: please report the receipt of any parcels in poor conditions to the TPS organiser within *[specify, e.g. 24 hours]* **of receipt, so that an alternative can be provided.**

2) Instructions for the storage of the samples

IMPORTANT: Upon receipt, the samples and controls must be stored at *[explain in detail the conditions for the storage of the samples and controls, e.g., -20 °C (frozen), -80 °C, +4 °C, room temperature, etc.]*, **until the analyses are performed.**

3) Instructions for the analyses

IMPORTANT: For each test, it is very important to strictly follow the recommendations given in the **TPS Technical Sheet that was provided on** *[provide the date]*.

The reprocessing of the samples before the analyses are detailed in the Appendices *[enter the number of appropriate Appendix/Appendices]* of the TPS Technical Sheet.

The details of the protocols for each test that will have to be implement by the participating laboratory are provided in the Appendices *[enter the number of appropriate Appendix/Appendices]* of the TPS Technical Sheet.

IMPORTANT: We remind the participating laboratory that for the benefit of the project and to obtain enough data, each participating laboratory has to evaluate all of the tests related to the methods for which it is registered in the TPS.

[if relevant a summary of the methods to be performed by each laboratory (identified by its panel code) can be provided in, e.g., the form of a table].

Each sample/control must be analysed in *[enter the number of replicates, e.g., single, duplicate, triplicate]* *[specify for each method/ test separately, if needed.]*)

The analyses must be performed **no later than** *[add the date].*

4) Special precautions

When carrying out the tests, the participating laboratory will have to make the necessary arrangements to prevent the risk of unintentional release of plant pathogens that they might be handling.

After performing the tests:
[specify; For example:
-any remainder of samples and controls are to be destroyed by autoclaving or by any other efficient means for inactivating viruses;
-consumables in contact with the samples or controls are to be autoclaved (tubes, microtitre plates, PCR plates, etc.);
-materials in contact with samples and controls are to be disinfected.]

5) Instructions for entering and submitting the results

The results must be entered in the TPS Results Form that is sent by e-mail. It is an *[specify, e.g., Excel file consisting of nine different sheets: one per test corresponding to the results tables to be completed for each test performed by the participant, and one for details regarding the RNA extraction.]*

<u>In these sheets, you must provide:</u>
[specify, e.g.,
   - the starting date and ending date for performing the analyses;
   - the qualitative (and quantitative, when available) results of the controls;
   - the qualitative (and quantitative, when available) results of the samples (for ELISA and molecular tests, where the samples are analysed in duplicate, the results for each sample are to be given both per well and overall);
   - if relevant and appropriate, the mode of calculation and the value(s) of the threshold and/or the Cq cut-off;
   -any comments concerning the obtaining of the results (e.g., the need to repeat the analysis, cross contamination, etc.);
   - the consumables used;
   - the equipment used;
   -any deviation from the protocols described in the TPS Technical Sheet;
   -any remarks, comments, difficulties encountered when carrying out the tests.]

This information is hugely important to contextualise the implementation of the TPS and to interpret the results in an appropriate manner.

The participating laboratory will have to **return the TPS Results Form as** *[specify, e.g.,* **an Excel file and as signed scans of printed files** by e-mail] to the following addresses: *[add e-mail address 1 of the TPS organiser]* **and** *[add e-mail address 2 of the TPS organiser]* **on** *[provide the date]* **at the latest.**

The final report will be transmitted to the participating laboratory on *[provide the date].*

Validated on *[add the date]*

The TPS organiser *[add the name and signature of responsible person for the TPS organisation]*	

Appendix 6. Example of Acknowledgement of Receipt of the Panel of Samples

[add logo]	**Acknowledgement of Receipt** **of the panel of samples**

To be returned <u>**within 24 hours of receiving**</u> the samples by e-mail to:
[add e-mail address 1 of the TPS organiser]
and
[add e-mail address 2 of the TPS organiser]

Identification of the TPS

<u>Pest name:</u> Tomato spotted wilt tospovirus
<u>TPS code:</u> TSWV

Identification of the participating laboratory

<u>Name of the laboratory:</u>
<u>Panel code (ID of the participant)[3]:</u>

- Our laboratory certifies the receipt from the TPS organiser of   panel(s) *[please specify the number]* consisting each of 22 samples numbered from S-1 to S-22, and the controls listed below (please tick the boxes according to what was received).

Sample/ control panel	Number of samples	NC	PC	NIC	PIC	PAC	Comments
				Content of the panel			
ELISA 1 ☐	22 ☐	1 ☐	1 ☐	/	/	/	
ELISA 2 ☐	22 ☐	1 ☐	1 ☐	/	/	/	
LFD 1 ☐	22 ☐	1 ☐	1 ☐	/	/	/	
LFD 2 ☐	22 ☐	1 ☐	1 ☐	/	/	/	
Molecular ☐	22 ☐	/	/	1 ☐	1 ☐	1 ☐	

- Date of receipt:_____

- I can confirm that the sample and control panels received are in line with the signed contract (TPS Participant contract signed on _____ *[please enter the date and tick the relevant box below*):
  - ☐ yes
  - ☐ no

    If «No»: Please specify (e.g., which sample panel is missing):
    _____
    _____

---

[3] *The panel code is in the form LXX, where XX is a two-digit number. It is written on the packaging of the parcel of samples. This code will be used for the presentation of the results in the test report to maintain confidentiality.*

- Our laboratory certifies that when received, the TPS materials were on dry ice and were then immediately stored at $\leq$-15 °C [*please tick the box*]:
  - ☐ yes
  - ☐ no
    - If «No»: Please detail any problems encountered:

    _____
    _____

- The sample condition upon receipt of the parcel was considered to be [*please tick the box*]:
  - ☐ Satisfactory
  - ☐ Not satisfactory
    - If «Not satisfactory»: Please detail the damaged and/or precise problem(s) encountered:

    _____
    _____
    _____

- Our laboratory certifies the receipt of the primer mix (PP) for conventional RT-PCR and primer probe mixes (PPS) for real-time RT-PCRs [*please tick the boxes if received*]:

Reagents (✓ tubes received)		Comments
Conventional RT-PCR	PP ☐	
Real-time RT-PCR_1	PPS ☐	
Real-time RT-PCR_2	PPS ☐	
Real-time RT-PCR_3	PPS ☐	

- Our laboratory also certifies the receipt of the following documents [*please tick the boxes if received*]:
  - ☐ the TPS TSWV Technical Sheet (received when registering)
  - ☐ the TPS TSWV Instruction Sheet
  - ☐ the TPS TSWV Results Form

Date   Name of the TPS participant   Signature

_____   _____   _____

Appendix 7. TPS Results Form

TPS Results Form (ELISA example)

[add logo]

TPS code [ ]   **Target organism** [ ]

Method   ELISA 1   **Test** [ ]

Identification of the participating laboratory	
Name of the laboratory	
Panel code	

Implementation of the test	
Start date for performing the analysis	
End date for performing the analysis	

Details about running/analysis of the test			
Plate reader	**Brand**	**Model**	**Wavelength of filters**

Specify any deviation(s) from the ELISA 1 protocols described in the TPS Technical Sheet
Remarks, comments, difficulties encountered when carrying out the ELISA 1 test

Results obtained from the controls

The ELISA test should be interpreted as instructed in the TPS Technical Sheet.

Controls	Repetition (well)	Absorbance value (OD) per well after 60 min	Average absorbance value (OD) per control after 60 min	Comments (if necessary)
Negative control (NC)-kit	1			
	2			
Negative control (NC)	1			
	2			
	3			
	4			

	5			
	6			
Buffer control (BC)	1			
	2			
	3			
	4			
Positive control (PC)-kit	1			
	2			
Positive control (PC)	1			
	2			
Other *(detail)*:		1		
		2		
Other *(detail)*:		1		
		2		

Determination of the positivity threshold

Mode of determination of the threshold (specify only if not carried out as recommended in the TPS Technical Sheet; include the reason)	
Value of the threshold	

Results obtained from the samples

The ELISA test should be interpreted as instructed in the TPS Technical Sheet.

Sample reference: ELISA 1	Repetition (well)	Absorbance value (OD) per well after 60 min	Average absorbance value (OD) per sample after 60 min	Qualitative results	Comments (if necessary)
S-1	1				
	2				
S-2	1				
	2				
S-3	1				
	2				
.....	1				
	2				

Validation by the TPS participant

Date		Name	
		Signature	

TPS Results Form (real-time PCR example)

[add logo]

TPS code [   ]   **Target organism** [   ]

Method [real-time PCR]   **Test** [   ]

Identification of the participating laboratory	
Name of the laboratory	
Panel code	

Implementation of the test	
Start date for performing the analysis	
End date for performing the analysis	

Details about running/analysis of the test		
	Supplier	**Name and reference**
Amplification kit		

	Brand	**Model**
Real-time PCR cycler		

Specify any deviation(s) from the real-time PCR protocol described in the TPS Technical Sheet (e.g., amplification conditions)

Remarks, comments, difficulties encountered when carrying out the real-time PCR protocol

Results obtained from the controls					
Controls	**Repetition (well)**	**Cq value per well**	**Average Cq value per control**	**Qualitative results**	**Comments (if necessary)**
Negative isolation control (NIC)	1				
	2				
Positive isolation control (PIC)	1				
	2				
Negative amplification control (NAC)	1				
	2				
	1				

Positive amplification control (PAC)		2				
Other *(detail)*:		1				
		2				
Other *(detail)*:		1				
		2				

Determination of the threshold				
Determination of the threshold		*If manual, detail the decision-making rules:*		
Value of the threshold				
If necessary: value of the Cq cut-off				

Results obtained from the samples					
Sample reference: molecular tests	Repetition (well)	Cq value per well	Average Cq value per sample	Qualitative results	Comments (if necessary)
S-1	1				
	2				
S-2	1				
	2				
S-3	1				
	2				
.....	1				
	2				

N.B. Please attach a copy of the real-time PCR report with amplification curves!

Validation by the TPS participant			
Date		Name	
		Signature	

Appendix 8. Example of the Outline of the TPS Report

Contents

1   Context of application ....{}
2   Disease and pathogen ....{}
3   Methodology of evaluation ....{}
  3.1   Common rules for selection of tests for TPS...{}
    3.1.1   Definition of the scope of testing...{}
    3.1.2   Weighting and targeted values for each criterion to be reached by a test......{}
    3.1.3   Collection of available data..{}
    3.1.4   Analysis of available data..{}
    3.1.5   First selection of tests..{}
    3.1.6   Preliminary studies...{}
    3.1.7   Selection of the final tests..{}
  3.2   Common rules for selection of participants for TPS ....{}
    3.2.1   Identification of potential participants for the TPS............................{}
    3.2.2   Weight and targeted values for each criterion to be reached by the participants..{}
    3.2.3   Sending invitations...{}
    3.2.4   Selection of the participants...{}
4   Preliminary study for evaluation of method performance ....{}
  4.1   Test selection..{}
  4.2   Material and methods..{}
  4.3   Results of preliminary study..{}
5   Tests performance study ....{}
  5.1   Evaluated parameters ....{}
  5.2   Test panel composition and preparation ....{}
  5.3   Assigned reference values...{}
  5.4   Homogeneity and stability testing ....{}
  5.5   Randomisation...{}
  5.6   Distribution of the samples ....{}
  5.7   Consumables ....{}
  5.8   Equipment and materials ....{}
  5.9   Methods ....{}
  5.10 Participants..{}
  5.11 Results ....{}
    5.11.1 Collected results...{}
    5.11.2 Outlier results ....{}
    5.11.3 Performance of individual tests...{}
    5.11.4 Comparison of the methods/tests and other diagnostic parameters ....{}
6   Conclusion: ....{}
References ....{}
Appendices..{}

References

Agstner B, Jones G (2020) VALITEST - D4.1 Report on stakeholder priorities for tests and general prioritization framework (Version 1). Zenodo:1–30. https://doi.org/10.5281/zenodo.5564619

Alič Š, Anthoine G, Chabirand A et al (2020) VALITEST - D1.1 Minimum performance parameters to select tests for validation and selection of laboratories for TPS (Version 2). Zenodo:1–30. https://doi.org/10.5281/zenodo.5561020

Anonymous (2008) International vocabulary of metrology – basic and general concepts and associated terms, 3rd edn. VIM

Anthoine G, Brittain I, Chabirand A, et al (2020) VALITEST - D1.3 List of tests for validation – Round 2 Validation of diagnostic tests to support plant health:1–16. https://doi.org/10.5281/zenodo.5561846

Aramburu J, Martí M (2003) The occurrence in north-east Spain of a variant of Tomato spotted wilt virus (TSWV) that breaks resistance in tomato (Lycopersicon esculentum) containing the Sw-5 gene. Plant Pathol 52:407. https://doi.org/10.1046/j.1365-3059.2003.00829.x

Boonham N, Smith P, Walsh K et al (2002) The detection of Tomato spotted wilt virus (TSWV) in individual thrips using real time fluorescent RT-PCR (TaqMan). J Virolog Methods 101(1-2):37–48. https://doi.org/10.1016/S0166-0934(01)00418-9. PMID: 11849682

Chabirand A, Loiseau M, Renaudin I, Poliakoff F (2017) Data processing of qualitative results from an interlaboratory comparison for the detection of "flavescence dorée" phytoplasma: How the use of statistics can improve the reliability of the method validation process in plant pathology. PLoS One 12:1–26. https://doi.org/10.1371/journal.pone.0175247

Chappé A-M, Chabirand A, Dahlin P et al (2020) VALITEST - D3.1 List of the criteria the reference materials have to meet for use in validation studies (Version 2). Zenodo:1–19. https://doi.org/10.5281/zenodo.5562337

Ciuffo M, Finetti-Sialer MM, Gallitelli D, Turina M (2005) First report in Italy of a resistance-breaking strain of Tomato spotted wilt virus infecting tomato cultivars carrying the Sw5 resistance gene. Plant Pathol 54:564. https://doi.org/10.1111/j.1365-3059.2005.01203.x

Commission Implementing Regulation (EU) 2019/2072 (2019, November 28) 10.12.2019

Debreczeni DE, Ruiz-Ruiz S, Aramburu J et al (2011) Detection, discrimination and absolute quantitation of Tomato spotted wilt virus isolates using real time RT-PCR with TaqMan ®MGB probes. J Virolog Methods 176:32–37. https://doi.org/10.1016/j.jviromet.2011.05.027

EFSA Panel on Plant Health (2012) Scientific Opinion on the pest categorisation of the tospoviruses. EFSA J 10:1–101. https://doi.org/10.2903/j.efsa.2012.2772

EPPO (2014) PM 7/122 (1) Guidelines for the organization of interlaboratory comparisons by plant pest diagnostic laboratories. EPPO Bull 44:390–399. https://doi.org/10.1111/epp.12162

© The Editor(s) (if applicable) and The Author(s) 2022

A. Vučurović et al. (eds.), *Critical Points for the Organisation of Test Performance Studies in Microbiology*, Plant Pathology in the 21st Century 12, https://doi.org/10.1007/978-3-030-99811-0

EPPO (2018a) PM 7/76 (5) Use of EPPO Diagnostic Standards. EPPO Bull 48:373–377. https://doi.org/10.1111/epp.12506

EPPO (2018b) PM 7/84 (2) Basic requirements for quality management in plant pest diagnostic laboratories. EPPO Bull 48:378–386. https://doi.org/10.1111/epp.12507

EPPO (2019) PM 7/98 (4) Specific requirements for laboratories preparing accreditation for a plant pest diagnostic activity. EPPO Bull 49:530–563. https://doi.org/10.1111/epp.12629

EPPO (2021a) PM 7/98 (5) Specific requirements for laboratories preparing accreditation for a plant pest diagnostic activity. EPPO Bull 51:468–498. https://doi.org/10.1111/epp.12780

EPPO (2021b) PM 7/147 (1) Guidelines for the production of biological reference material. EPPO Bull 51:499–506. https://doi.org/10.1111/epp.12781

Finneti-Sialer MMF, Lanave C, Padula M et al (2002) Occurrence of two distinct tomato spotted wilt virus subgroups in southern Italy. J Plant Pathol 84:145–152. https://doi.org/10.4454/jpp.v84i3.1100

Hassani-Mehraban A, Westenberg M, Verhoeven JTJ et al (2016) Generic RT-PCR tests for detection and identification of tospoviruses. J Virolog Methods 233:89–96. https://doi.org/10.1016/j.jviromet.2016.03.015

IPPC (2021) Strategic Framework for the International Plant Protection Convention (IPPC) 2020–2030. Protecting global plant resources and facilitating safe trade. FAO behalf Secr Int Plant Prot Conv.

ISO/IEC 17025 (2005) Testing and Calibration Laboratories. International Organisation for Standardisation/International Electrotechnical Committee, Geneva

ISO/IEC 17043 (2010) Conformity assessment—General requirements for proficiency testing. International Organisation for Standardisation/International Electrotechnical Committee, Geneva

ISO/IEC 30 (2015) Reference materials—Selected terms and definitions. International Organisation for Standardisation/International Electrotechnical Committee, Geneva

ISO/IEC 13528 (2015) Statistical methods for use in proficiency testing by interlaboratory comparison. International Organisation for Standardisation/International Electrotechnical Committee, Geneva

ISO/IEC 17034 (2016) General requirements for the competence of reference material producers. International Organisation for Standardisation/International Electrotechnical Committee, Geneva

Langton SD, Chevennement R, Nagelkerke N, Lombard B (2002) Analysing collaborative trials for qualitative microbiological methods: accordance and concordance. Int J Food Microbiol 79:175–181

Massart S., Lebas B, Chabirand A, et al (2022) Guidelines for improving analyses of validation datasets for plant health diagnostic tests. EPPO Bull. In preparation

Mortimer-Jones SM, Jones MGK, Jones RAC et al (2009) A single tube, quantitative real-time RT-PCR assay that detects four potato viruses simultaneously. J Virolog Methods 161:289–296. https://doi.org/10.1016/j.jviromet.2009.06.027

Mumford RA, Barker I, Wood KR (1994) The detection of tomato spotted wilt virus using the polymerase chain reaction. J Virolog Methods 46:303–311. https://doi.org/10.1016/0166-0934(94)90002-7

Panno S, Davino S, Rubio L et al (2012) Simultaneous detection of the seven main tomato-infecting RNA viruses by two multiplex reverse transcription polymerase chain reactions. J Virolog Methods 186:152–156. https://doi.org/10.1016/j.jviromet.2012.08.003

Petter F, Trontin C, Gianinazzi C (2021) VALITEST - D6.9 Final Plan for the Dissemination and Exploitation of Project Results (PDER) (Version 1). Zenodo:1–61. https://doi.org/10.5281/zenodo.5653347

Roberts CA, Dietzgen RG, Heelan LA, MaclLean DJ (2000) Real-time RT-PCR fluorescent detection of tomato spotted wilt virus. J Virolog Methods 88:1–8. https://doi.org/10.1016/S0166-0934(00)00156-7

Rodrigues SM, Demokritou P, Dokoozlian N et al (2017) Nanotechnology for sustainable food production: promising opportunities and scientific challenges. Envir Sci Nano 4:767–781. https://doi.org/10.1039/c6en00573j

Rybicki EP (2015) A Top Ten list for economically important plant viruses. Arch Virol 160:17–20. https://doi.org/10.1007/s00705-014-2295-9

Scholthof KBG, Adkins S, Czosnek H et al (2011) Top 10 plant viruses in molecular plant pathology. Mol Plant Pathol 12:938–954. https://doi.org/10.1111/j.1364-3703.2011.00752.x

Trontin C, Agstner B, Altenbach D et al (2021) VALITEST: Validation of diagnostic tests to support plant health. EPPO Bull 51:198–206. https://doi.org/10.1111/epp.12740

Turina M, Tavella L, Ciuffo M (2012) Chapter 12 - Tospoviruses in the mediterranean area. In: Loebenstein G, HBT-A L (eds) VR Viruses and virus diseases of vegetables in the mediterranean basin. Academic Press, pp 403–437

Vučurović A, Bulajić A, Stanković I et al (2012) Non-persistently aphid-borne viruses infecting pumpkin and squash in Serbia and partial characterization of Zucchini yellow mosaic virus isolates. Eur J Plant Pathol 133:935–947. https://doi.org/10.1007/s10658-012-9964-x

Zarzyńska-Nowak A, Hasiów-Jaroszewska B, Korbecka-Glinka G et al (2018) A multiplex RT-PCR assay for simultaneous detection of Tomato spotted wilt virus and Tomato yellow ring virus in tomato plants. Canadian J Plant Pathol 40:580–586. https://doi.org/10.1080/07060661.2018.1503195

Printed in the United States
by Baker & Taylor Publisher Services